Transmission Electron Microscopy

A Textbook for Materials Science

Transmission Electron Microscopy

A Textbook for Materials Science

David B. Williams

Lehigh University
Bethlehem, Pennsylvania

and

C. Barry Carter

University of Minnesota
Minneapolis, Minnesota

Plenum Press • New York and London

Library of Congress Cataloging in Publication Data

Williams, David B. (David Bernard), 1949–
Transmission electron microscopy: a textbook for materials science / David B. Williams and C. Barry Carter.
p. cm.
Includes bibliographical references and index.
ISBN 0-306-45247-2 (hardbound).—ISBN 0-306-45324-X (pbk.)
1. Materials—Microscopy. 2. Transmission electron microscopy. I. Carter, C. Barry. II. Title.
TA417.23.W56 1996 96-28435
502'.8'25—dc20 CIP

ISBN 0-306-45247-2 (Hardbound)
ISBN 0-306-45324-X (Paperback)

A Division of Plenum Publishing Corporation
233 Spring Street, New York, N. Y. 10013

10 9 8 7 6 5 4 3 2 1

Printed in the United States of America

To our parents,
Walter Dennis and Mary Isabel Carter,
and Joseph and Catherine Williams,
who made everything possible

Foreword

Electron microscopy has revolutionized our understanding of materials by completing the *processing–structure–properties* links down to atomistic levels. It now is even possible to tailor the microstructure (and mesostructure) of materials to achieve specific sets of properties; the extraordinary abilities of modern transmission electron microscopy—TEM—instruments to provide almost all of the structural, phase, and crystallographic data allow us to accomplish this feat. Therefore, it is obvious that any curriculum in modern materials education must include suitable courses in electron microscopy. It is also essential that suitable texts be available for the preparation of the students and researchers who must carry out electron microscopy properly and quantitatively.

The 40 chapters of this new text by Barry Carter and David Williams (like many of us, well schooled in microscopy at Cambridge and Oxford) do just that. If you want to learn about electron microscopy from specimen preparation (the ultimate limitation); or via the instrument; or how to use TEM correctly to perform imaging, diffraction, and spectroscopy—it's all there! This is, to my knowledge, the only complete text now available that includes all the remarkable advances made in the field of TEM in the past 30 to 40 years. The timing for this book is just right and, personally, it is exciting to have been part of the developments it covers—developments that have impacted so heavily on materials science.

In case there are people out there who still think TEM is just taking pretty pictures to fill up one's bibliography, please stop, pause, take a look at this book, and digest the extraordinary intellectual demands required of the microscopist in order to do the job properly: crystallography, diffraction, image contrast, inelastic scattering events, and spectroscopy. Remember, these used to be fields in themselves. Today, one has to understand the fundamentals of *all* of these areas before one can hope to tackle significant problems in materials science. TEM is a technique of characterizing materials down to the atomic limits. It must be used with care and attention, in many cases involving teams of experts from different venues. The fundamentals are, of course, based in physics, so aspiring materials scientists would be well advised to have prior exposure to, for example, solid-state physics, crystallography, and crystal defects, as well as a basic understanding of materials science, for without the latter how can a person see where TEM can (or may) be put to best use?

So much for the philosophy. This fine new book definitely fills a gap. It provides a sound basis for research workers and graduate students interested in exploring those aspects of structure, especially defects, that control properties. Even undergraduates are now expected (and rightly) to know the basis for electron microscopy, and this book, or appropriate parts of it, can also be utilized for undergraduate curricula in science and engineering.

The authors can be proud of an enormous task, very well done.

G. Thomas
Berkeley, California

Preface

How is this book any different from the many other books that deal with TEM? It has several unique features, but the most distinguishing one, we believe, is that it can really be described as a "textbook"—that is, one designed to be used primarily in the classroom rather than in the research laboratory. We have constructed the book as a series of relatively small chapters (with a few notable exceptions!). The contents of many chapters can be covered in a typical lecture of 50 to 75 minutes. The style is informal for easier reading; it resembles an oral lecture rather than the formal writing you would encounter when reading research papers.

In our experience, the TEM books currently available fall into three major categories. They may be too theoretical for many materials science students; they attempt to cover all kinds of electron microscopy in one volume, which makes it difficult to include sufficient theory on any one technique; or they are limited in the TEM topics they cover. The rapid development of the TEM field has meant that many of the earlier books must automatically be placed in the third category. Although these books are often invaluable in teaching, we have not found them generally suitable as the course textbook in a senior-year undergraduate or first-year graduate course introducing TEM, so we have endeavored to fill this perceived gap.

Since this text is an introduction to the whole subject of TEM, we incorporate *all* aspects of a modern TEM into an integrated whole. So, rather than separating out the broad-beam and convergent-beam aspects of the subject (the traditional structural analysis or imaging versus the "chemical" analysis or "new" techniques), we treat these two aspects as different sides of the same coin. Thus scanning-beam (STEM) imaging is just another way to form an image in a TEM. There is no reason to regard "conventional" bright-field and "conventional" dark-field imaging as any more fundamental ways of imaging the specimen than annular dark-field imaging—or even secondary-electron or STEM *Z*-contrast modes. Similarly, convergent-beam and scanning-beam diffraction are integral parts of electron diffraction, and are complementary to selected-area diffraction. Inelastic electron scattering is the source of both Kikuchi lines and characteristic X-rays. So we don't deliberately split off "conventional" microscopy from "analytical" microscopy.

Our approach is to thread two fundamental questions throughout the text.

Why should we use a particular technique?

How do we put the idea into practice?

We attempt to establish a sound theoretical basis where necessary, although not always giving all the details. We then use this knowledge to build a solid understanding of how we use the instrument. The text is illustrated with examples from across the fields of materials science and engineering and, where possible, a sense of the history of the technique is introduced. We keep references to a minimum and generally accepted concepts are not specifically credited, although numerous classical general references are included.

We both have extensive teaching and research backgrounds in all aspects of TEM comprising diffraction, imaging, and microanalysis. Our research in TEM of materials spans metals, ceramics, composites, and semiconductors. We each bring more than 25 years of TEM experience to the book, and have contributed to the training of a generation of (we hope) skilled electron microscopists. We found that writing the book broadened our own knowledge considerably and was actually fun on some occasions. We hope you experience the same reactions after reading it.

Lastly, we encourage you to send us any comments (positive or negative). We can both be reached by email: dbwl @lehigh.edu and carter@cems.umn.edu

Acknowledgments

We have spent the best part of a decade in the conception and gestation of this text and such an endeavor can't be accomplished in isolation. Our first acknowledgments must be to our wives, Margie and Bryony, and our families, who have borne the brunt of our absences from home (and occasionally the brunt of our presence, also!).

We have both been fortunate to work with other microscopists, post-doctoral associates, and graduate students who have taught us much and contributed significantly to the examples in the text. We would like to thank a few of these colleagues directly: Dave Ackland, Ian Anderson, Charlie Betz, John Bruley, Dov Cohen, Ray Coles, Vinayak Dravid, Joe Goldstein, Brian Hebert, Jason Heffelfinger, John Hunt, Matt Johnson, Vicki Keast, Ron Liu, Charlie Lyman, Stuart McKernan, Joe Michael, Grant Norton, Sundar Ramamurthy, René Rasmussen, Kathy Repa, Al Romig, David A. Smith, Changmo Sung, Caroline Swanson, Ken Vecchio, and Mike Zemyan.

In addition, many other colleagues and friends in the fields of microscopy and microanalysis have helped with the book (even if they weren't aware of it). These include: Ron Anderson, Jim Bentley, Geoff Campbell, Graham Cliff, David Cockayne, the late Chuck Fiori, Peter Goodhew, Ron Gronsky, Peter Hawkes, David Joy, Roar Kilaas, Gordon Lorimer, Harald Müllejans, Dale Newbury, Mike O'Keefe, John Steeds, Peter Swann, Gareth Thomas, Patrick Veyssière, Nestor Zaluzec, and Elmar Zeitler. In addition, many other microscopists kindly provided the figures that we acknowledge individually in the list at the end of the book.

We have received financial support for our microscopy studies through several federal agencies; without this support none of the research that underpins the contents of this book would have been accomplished. In particular, DBW wishes to acknowledge the National Science Foundation (Division of Materials Research) for almost 20 years of continuous funding, the National Aeronautics and Space Administration, the Department of Energy (Basic Energy Sciences), Sandia National Laboratories, and the Materials Research Center at Lehigh, which supports the microscopy laboratory. Portions of the text were written while DBW was on sabbatical or during extended visits to Chalmers University, Göteborg, with Gordon Dunlop and Hans Nordén; the Max Planck Institut für Metallforschung, Stuttgart, with Manfred Rühle; and Los Alamos National Laboratory, with Terry Mitchell. CBC wishes to acknowledge the Department of Energy (Basic Energy Sciences), the National Science Foundation (Division of Materials Research), the Center for Interfacial Engineering at the University of Minnesota, the Materials Science Center at Cornell University, and the SHaRE Program at Oak Ridge National Laboratories. This text was started while CBC was with the Department of Materials Science and Engineering at Cornell University.

Despite our common scientific beginnings as undergraduates in Christ's College, Cambridge, we learned our trade under different microscopists; DBW with Jeff Edington in Cambridge and CBC with Sir Peter Hirsch and Mike Whelan in Oxford. Not surprisingly, the classical texts by these renowned microscopists are referred to throughout this book. They influenced our own views of TEM tremendously, unavoidably contributing to any bias in our opinions, notation, and approach to the whole subject.

List of Acronyms

The field of TEM is a rich source of acronyms, behind which we hide both simple and esoteric concepts. While the generation of new acronyms can be a source of original thinking (e.g., see ALCHEMI), it undoubtedly makes for easier communication in many cases and certainly reduces the length of voluminous textbooks. You have to master this strange language before being accepted into the community of microscopists, so we present a comprehensive listing that you should memorize.

ACF absorption correction factor
A/D analog to digital (converter)
ADF annular dark field
AEM analytical electron microscope/microscopy
AES Auger electron spectrometer/spectroscopy
AFF aberration-free focus
ALCHEMI atom location by channeling-enhanced microanalysis
ANL Argonne National Laboratory
APB anti-phase domain boundary
ASU Arizona State University
ATW atmospheric thin window

BF bright field
BFP back focal plane
BSE backscattered electron
BSED backscattered-electron diffraction
BZB Brillouin-zone boundary

C(1, 2, etc.) condenser (1, 2, etc.) lens
CB coherent bremsstrahlung
CBED convergent-beam electron diffraction
CBIM convergent-beam imaging
CCD charge-coupled device
CCF cross-correlation function
CCM charge-collection microscopy
CDF centered dark field
CF coherent Fresnel/Foucault
CFEG cold field-emission gun
CL cathodoluminescence
CRT cathode-ray tube
CS crystallographic shear
CSL coincident-site lattice

DF dark field
DOS density of states
DP diffraction pattern
DQE detection quantum efficiency
DSTEM dedicated scanning transmission electron microscope/microscopy
DTSA desktop spectrum analyzer

EBIC electron beam-induced current/conductivity
EELS electron energy-loss spectrometry
EFI energy-filtered imaging
ELNES energy-loss near-edge structure
ELP energy-loss program (Gatan)
EMMA electron microscope microanalyzer
EMS electron microscopy image simulation
EPMA electron probe microanalyzer
ESCA electron spectroscopy for chemical analysis
ESI electron spectroscopic imaging
EXAFS extended X-ray absorption fine structure
EXELFS extended energy-loss fine structure

FCF fluorescence correction factor
FEG field-emission gun
FET field-effect transistor
FFT fast Fourier transform

FOLZ first-order Laue zone
FSE fast secondary electron
FTP file transfer protocol
FWHM full width at half maximum
FWTM full width at tenth maximum

GB grain boundary
GCS generalized cross section
GIF Gatan image filter
GOS generalized oscillator strength

HAADF high-angle annular dark field
HOLZ higher-order Laue zone
HPGe high-purity germanium
HRTEM high-resolution transmission electron microscope/microscopy
HV high vacuum
HVEM high voltage electron microscope/microscopy

IDB inversion domain boundary
IEEE International Electronics and Electrical Engineering
IG intrinsic Ge
IVEM intermediate voltage electron microscope/microscopy

K–M Kossel–Möllenstedt

LEED low-energy electron diffraction
LLS linear least-squares
LUT look-up table

MC minimum contrast
MCA multichannel analyzer
MDM minimum detectable mass
MLS multiple least-squares
MMF minimum mass fraction
MSDS material safety data sheets

NCEMSS National Center for Electron Microscopy simulation system
NIH National Institutes of Health
NIST National Institute of Standards and Technology

OR orientation relationship
OTEDP oblique-textured electron diffraction pattern

PB phase boundary
P/B peak-to-background ratio
PEELS parallel electron energy-loss spectrometer/spectrometry
PIMS Precision Ion-Milling System™
PIPS Precision Ion-Polishing System™
PM photomultiplier
POA phase-object approximation

QHRTEM quantitative high-resolution transmission electron microscopy

RB translation boundary (yes, it does!)
RCP rocking-beam channeling patterns
RDF radial distribution function
REM reflection electron microscope/microscopy
RHEED reflection high-energy electron diffraction
RHF relativistic Hartree–Fock
RHFS relativistic Hartree–Fock–Slater

SAD selected-area diffraction
SE secondary electron
SEELS serial electron energy-loss spectrometer/spectrometry
SEM scanning electron microscope/microscopy
SF stacking fault
SHRLI simulated high-resolution lattice images
SIMS secondary ion mass spectrometry
S/N signal-to-noise ratio
SOLZ second-order Laue zone
SRM standard reference material
STEM scanning transmission electron microscope/microscopy
STM scanning tunneling microscope/microscopy

TB twin boundary
TEM transmission electron microscope/microscopy
TMBA too many bloody acronyms

UHV ultrahigh vacuum
UTW ultrathin window

V/F voltage to frequency (converter)
VLM visible-light microscope/microscopy

WB weak beam
WBDF weak-beam dark field
WDS wavelength-dispersive spectrometer/spectrometry
WP whole pattern
WPOA weak-phase object approximation
WWW World Wide Web

XANES X-ray absorption near-edge structure
XEDS X-ray energy-dispersive spectrometer/spectrometry
XRD X-ray diffraction

YBCO yttrium–barium–copper oxide
YAG yttrium–aluminum garnet

ZAF atomic number, absorption, fluorescence (correction)
ZAP zone-axis pattern
ZOLZ zero-order Laue zone

List of Symbols

We use a large number of symbols. Because we are constrained by the limits of our own and the Greek alphabets we often use the same symbol for different terms, which can confuse the unwary. We have tried to be consistent where possible but undoubtedly we have not always succeeded. The following (not totally inclusive) list may help if you remain confused after reading the text.

a relative transition probability
a_0 Bohr radius
a, b, c lattice vectors
a*, b*, c* reciprocal lattice vectors
A amplitude of scattered beam
A amperes
A absorption correction factor
A active area of the detector
A Richardson's constant
A atomic weight
Å Ångstrom
$\mathcal{A}$ Bloch wave amplitude
$A(\mathbf{U})$ aperture function

b beam-broadening parameter
$\mathbf{b}_e$ edge component of the Burgers vector
$\mathbf{b}_p$ Burgers vector of partial dislocation
$\mathbf{b}_T$ Burgers vector of total dislocation
B beam direction
B magnetic field strength
B background intensity
$B(\mathbf{U})$ aberration function

c velocity of light
C contrast
C composition
C_a astigmatism aberration coefficient
C_c chromatic aberration coefficient
$C_\mathbf{g}$ **g** component of Bloch wave
C_s spherical aberration coefficient
C_X fraction of X atoms on specific sites
$C_\mathbf{0}$ amplitude of direct beam
C_ε combination of the elastic constants
$(C_s\lambda)^{1/2}$ scherzer
$(C_s\lambda^3)^{1/4}$ glaser

d beam (probe) diameter
d diameter of spectrometer entrance aperture
d spacing of moiré fringes
d interplanar spacing
d_c effective source size
d_d diffraction-limited beam diameter
d_g Gaussian beam diameter
$d_{hk\ell}$ $hk\ell$ interplanar spacing
d_{im} smallest resolvable image distance
d_{ob} smallest resolvable object distance
d_s spherical-aberration limited beam diameter
d_{eff} effective entrance aperture diameter at recording plane
dz thickness of a diffracting slice
$d\sigma/d\Omega$ differential cross section of one atom
D change in focus
D distance from projector crossover to recording plane
D electron dose
D_{im} depth of focus
D_{ob} depth of field
D_1, D_2 tie-line points on dispersion surfaces in presence of defect

e charge on the electron
E energy
$\mathbf{E}$ electric field
E Young's modulus
E_a spatial coherence envelope
E_c chromatic aberration envelope
E_c critical ionization energy
E_d displacement energy
E_K ionization energy for K-shell electron
E_L ionization energy for L-shell electron
$\mathcal{E}$ total energy
$\mathcal{E}$ energy loss
$\mathcal{E}_m$ average energy loss
$\mathcal{E}_P$ plasmon energy loss
E_P plasmon energy
E_t threshold energy
E_0 beam energy
$E(\mathbf{U})$ envelope function
$E_c(\mathbf{u})$ envelope function for chromatic aberration
$E_d(\mathbf{u})$ envelope function for specimen drift
$E_D(\mathbf{u})$ envelope function for the detector
$E_s(\mathbf{u})$ envelope function for the source
$E_v(\mathbf{u})$ envelope function for specimen vibration

f focal length
$f(\mathbf{r})$ strength of object at point x,y
$f(\theta)$ atomic scattering factor/amplitude
f_x scattering factor for X-rays
$f_i(x)$ residual of least-squares fit
F Fano factor
F fluorescence correction factor
$\mathbf{F}$ Lorentz force
F_B fraction of B alloying element
$F_{\mathbf{g}}$ special value of $F(\theta)$ when θ is the Bragg angle
$F(P)$ Fourier transform of plasmon intensity
$F(\mathbf{u})$ Fourier transform of $f(\mathbf{r})$
$F(0)$ Fourier transform of elastic intensity
$F(1)$ Fourier transform of single-scattering intensity
$F(\theta)$ structure factor

$\mathbf{g}$ diffraction vector (magnitude of $\mathbf{K}$ at the Bragg angle)
$g(\mathbf{r})$ intensity of image at point (x,y)
G Bragg reflection
G radius of a HOLZ ring
G giga
$G(\mathbf{u})$ Fourier transform of g($\mathbf{r}$)

h Planck's constant
h distance from specimen to the aperture
h(r) contrast transfer function
$(hk\ell)$ Miller indices of a crystal plane
$hk\ell$ indices of diffraction spots from $hk\ell$ plane

H spacing of the reciprocal-lattice planes parallel to the electron beam
$H(\mathbf{u})$ Fourier transform of $h(\mathbf{r})$

I intensity
I intrinsic line width of the detector
i_e emission current
i_f filament heating current
$I_{\mathbf{g}}$ intensity in the diffracted beam
I_K K-shell intensity above background
$I(\mathbf{k})$ kinematical intensity
I_P intensity in the first plasmon peak
I_T total transmitted intensity
I_0 intensity in the zero-loss peak
$I_{\mathbf{0}}$ intensity in the direct beam
$I(\ell)$ low-loss spectrum intensity

J current density

k magnitude of the wave vector
k Boltzmann's constant
k kilo
$\mathbf{k}_I$ $\mathbf{k}$-vector of the incident wave
$\mathbf{k}_D$ $\mathbf{k}$-vector of the diffracted wave
k_{AB} Cliff–Lorimer factor
K bulk modulus
K Kelvin
K Kramers constant
K sensitivity factor
K inner core shell/characteristic X-ray line/ionization edge
$\mathbf{K}$ change in $\mathbf{k}$ due to diffraction
$\mathbf{K}_B$ magnitude of $\mathbf{K}$ at the Bragg angle
K_o kernel

L camera length

m number of focal increments
m_0 rest mass of the electron
M magnification
M mega
M_A angular magnification
M_T transverse magnification
M_1, M_2 tie-line points on dispersion surfaces

n integer
n free-electron density
n nano
$\mathbf{n}$ vector normal to the surface
n_s number of electrons in the ionized subshell
N noise
N number of counts
N number of atoms per unit area
N $h + k + \ell$

$N(E)$ number of bremsstrahlung photons of energy E
N_0 Avogadro's number

O direct beam

p pico
p momentum
p integer
P peak intensity
P FWHM of a randomized electronic pulse generator
P_K probability of K-shell ionization
$P(z)$ scattering matrix for a slice of thickness z

Q number of scattering events per unit distance
Q cross section

r radius
r distance a wave propagates
r power term to fit background in EELS spectrum
r_M image translation distance
$\mathbf{r}_n$ lattice vector
$\mathbf{r}^*$ reciprocal lattice vector
r_{ast} astigmatism disk radius
r_{chr} chromatic-aberration disk radius
r_{sph} spherical-aberration disk radius
r_{min} minimum disk radius
r_{th} theoretical disk radius
$\mathbf{r}'_n$ lattice vector in strained crystal
r_0 maximum radius of DP in focal plane of spectrometer
$\mathbf{R}$ crystal lattice vector
R count rate
R resolution of XEDS detector
R distance on screen between diffraction spots
$\mathbf{R}_n$ lattice displacement vector

$\mathbf{s}$ excitation error or deviation parameter
$\mathbf{s}_\mathbf{R}$ excitation error due to defect
$\mathbf{s}_z$ ($\mathbf{s}_g$) excitation error in the z direction
s_{eff} effective excitation error
S distance from the specimen to detector
S signal
S standard deviation for n measurements
sr steradians

$\mathbf{t}$ shift vector between the ZOLZ and the HOLZ
t' absorption path length
$T(\mathbf{u})$ objective-lens transfer function
$T_{eff}(\mathbf{u})$ effective transfer function

u object distance
$\mathbf{u}$ unit vector along the dislocation line
$\mathbf{u}^*$ vector normal to the ZOLZ
$u_\mathbf{k}$ displacement field
U overvoltage
$U_\mathbf{g}$ Fourier component of the perfect-crystal potential

v image distance
v velocity of an electron
V accelerating voltage
$\mathcal{V}$ potential energy
V_c the volume of the unit cell
V_c inner potential of cavity
V_t projected potential through the thickness of the specimen
$V(\mathbf{r})$ crystal inner potential

T absolute temperature
T Tesla
T_C period of rotation

(UVW) indices of a crystal direction
UVW indices of beam direction

w $s\xi_\mathbf{g}$ (excitation error × extinction distance)

× times
x distance
× times (magnification)
x, y, z atom coordinates
X FWHM due to detector

y parallax shift in the image
y displacement at the specimen

z specimen height (distance along the optic axis)
Z atomic number/atomic number correction factor

Greek symbols

α phase shift due to defect
α semiangle of incidence/convergence
α X-ray take-off angle
α_{opt} optimum convergence semiangle

β brightness
β ratio of electron velocity to light velocity
β semiangle of collection
β_{opt} optimum collection semiangle

γ degree of spatial coherence
γ phase of direct beam

Δ change/difference
Δ width of energy window
$\Delta\phi$ phase difference
$\Delta\theta_i$ angles between Kossel–Möllenstedt fringes
Δ_{AB} difference in mass-absorption coefficients
ΔE energy spread
$\Delta\mathcal{E}_P$ plasmon line width
Δf maximum difference in focus
Δf_{AFF} aberration-free (de)focus
Δf_{MC} minimum contrast defocus
Δf_{opt} optimum defocus

Δf_{sch} Scherzer defocus
Δh relative depth in specimen
ΔI change in intensity
Δp parallax shift
ΔV change in the inner potential
Δx path difference
Δx half-width of image of undissociated screw dislocation
Δx_{res} resolution at Scherzer defocus
Δz change in height
δ angle between detector normal and line from detector to specimen
δ diameter of disk image
δ diffuseness of interface
δ fluorescence enhancement ratio
δ precipitate/matrix misfit

ε angle of deflection
ε detector efficiency
ε energy to create an electron–hole pair
ε strain
ε_0 permittivity of free space (dielectric constant)

$\eta(\theta)$ phase of the atomic scattering factor
η phase change
ν Poisson's ratio

$\mathbf{\Phi}$ work function
$\mathbf{\Phi}_A^{\Delta\rho t}$ X-ray emission from element A in an isolated thin film
ϕ angle between Kikuchi line and diffraction spot
ϕ angle between two Kikuchi line pairs
ϕ angle between two planes
ϕ angle of tilt between stereo images
ϕ phase of a wave
ϕ^* complex conjugate of ϕ
$\phi_{\mathbf{g}}$ amplitude of the diffracted beam
$\phi_{\mathbf{0}}$ amplitude of the direct beam
ϕ_x angle of deflection of the beam
$\mathbf{\phi}(\rho t)$ depth distribution of X-ray production

χ wave vector outside the specimen
χ_G wave vector which terminates on the point G in reciprocal space
χ_O wave vector which terminates on the point O in reciprocal space
$\chi(\mathbf{u})$ phase-distortion function

κ thermal conductivity

$\xi_{\mathbf{g}}$ extinction distance for diffracted beam
$\xi_{\mathbf{g}}'$ absorption parameter
$\xi_{\mathbf{0}}$ extinction distance for direct beam
$\xi_{\mathbf{g}}^{abs}$ absorption-modified $\xi_{\mathbf{g}}$

λ mean-free path
λ wavelength
λ_c coherence length
λ_P plasmon mean free path
λ^{-1} radius of Ewald sphere

μ micro
μ refractive index
μ/ρ mass absorption coefficient
$\mu^{(j)}(\mathbf{r})$ Bloch function

ν frequency

ψ amplitude of a wave
ψ^T total wave function
ψ^{tot} total wave function

ρ angle between directions
ρ density
ρ_c information limit due to chromatic aberration
$\rho(\mathbf{r})$ radial distribution function
ρt mass thickness
ρ_i^2 area of a pixel

σ scattering cross section of one atom
σ standard deviation
σ stress
σ_K ionization cross section for K-shell electron
σ_T total scattering cross section
$\sigma_K(\beta\Delta)$ partial ionization cross section

θ scattering semiangle
θ_B Bragg angle
θ_C cut-off semiangle
θ_E characteristic scattering semiangle
θ_0 screening parameter

τ detector time constant
τ dwell time

ω fluorescence yield
ω_c cyclotron frequency
ω_p plasmon frequency
Ω solid angle of collection

$\otimes$ convolution (multiply and integrate)

Contents

IV ■ Spectrometry

32 ■ X-ray Spectrometry

33 ■ The XEDS–TEM Interface

34 ■ Qualitative X-ray Analysis

35 ■ Quantitative X-ray Microanalysis

36 ■ Spatial Resolution and Minimum Detectability

37 ■ Electron Energy-Loss Spectrometers

Transmission Electron Microscopy

A Textbook for Materials Science

Transmission Electron Microscopy

I Basics

The Transmission Electron Microscope

1

CHAPTER PREVIEW

A typical commercial transmission electron microscope (TEM) costs about $2 for each electron volt of energy in the beam, and if you add on all the options, it can cost about $4–5 per eV. As you'll see, we use beam energies in the range from 100,000–400,000 eV, so a TEM becomes an extremely expensive piece of equipment. Consequently, there have to be very sound scientific reasons for investing such a large amount of money in one microscope. In this chapter (which is just a brief overview of many of the concepts that we'll talk about in detail throughout the book) we start by introducing you to some of the historical development of the TEM because the history is intertwined with some of the reasons why you need to use a TEM to characterize materials. Other reasons for using TEM appeared as the instrument developed. Unfortunately, coupled with the advantages are some serious drawbacks, which limit the microscope performance, and you must be just as aware of the instrument's limitations as you are of its advantages, so we summarize these also.

A TEM can appear in several different forms, all of which are described by different acronyms such as HRTEM, STEM, and AEM, and we'll introduce you to these different instruments. We'll also use the same acronym to denote both the technique (microscopy) and the instrument (microscope). We regard all of the different types of TEM as simply variations on a basic theme and that is why only "TEM" is in the book title. We will describe some of the basic physical characteristics of the electron. Throughout the book you'll have to confront some physics and mathematics every now and again. The reason for this is because understanding what we can do with a TEM and why we operate it in certain ways is governed by the fundamental physics of electrons, how electrons are controlled by magnetic fields in the microscope, and how electrons interact with materials.

Finally, we will summarize some of the most popular computer software packages for TEM. We will refer to many of these throughout the text. We are including them in the first chapter to emphasize the role of the computer in today's TEM analysis.

The Transmission Electron Microscope

1

1.1. WHY USE ELECTRONS?

Why should we use an electron microscope? Historically, TEMs were developed because of the limited image resolution in light microscopes, which is imposed by the wavelength of visible light. Only after electron microscopes were developed was it realized that there are many other equally sound reasons for using electrons, most of which are utilized to some extent in a modern TEM. By way of introduction to the topic let's look at how the TEM developed and the pros and cons of using such an instrument.

1.1.A. An Extremely Brief History

Louis de Broglie (1925) first theorized that the electron had wave-like characteristics, with a wavelength substantially less than visible light. Then Davisson and Germer (1927) and Thompson and Reid (1927) independently carried out their classic electron diffraction experiments which demonstrated the wave nature of electrons. It didn't take long for the idea of an electron microscope to be proposed, and the term was first used in the paper of Knoll and Ruska (1932). In this paper they developed the idea of electron lenses into a practical reality, and demonstrated electron images taken on the instrument shown in Figure 1.1. This was a most crucial step, for which Ruska received the Nobel Prize, somewhat late, in 1986. Within a year of Knoll and Ruska's publication, the resolution limit of the light microscope was surpassed. Ruska, surprisingly, revealed that he hadn't heard of de Broglie's ideas about electron waves and thought that the wavelength limit didn't apply to electrons. TEMs were developed by commercial companies only four years later. The Metropolitan-Vickers EM1 was the first commercial TEM. It was built in the UK in 1936, but apparently it didn't work very well and regular production was really started by Siemens and Halske in Germany in 1939. TEMs became widely available from several other sources (Hitachi, JEOL, Philips and RCA, *inter alia*) after the conclusion of World War II.

For materials scientists a most important development took place in the late 1940s when Heidenreich (1949) first thinned metal foils to electron transparency. This work was followed up by Bollman in Switzerland and Hirsch and co-workers in Cambridge. Because so much of the early TEM work examined metal specimens, the word "foil" has come to be synonymous with "specimen." In addition, the Cambridge group also developed the theory of electron diffraction contrast with which we can now identify, often in a quantitative manner, all known line and planar crystal defects in TEM images. This theoretical work is summarized in a formidable but essential text often referred to as the "Bible" of TEM (Hirsch *et al.* 1977). For the materials scientist, practical applications of the TEM for the solution of materials problems were pioneered in the United States by Thomas and first clearly expounded in his text (Thomas 1962). Other materials-oriented texts followed, e.g., Edington (1976) and Thomas and Goringe (1979).

Today, TEMs constitute arguably the most efficient and versatile tools for the characterization of materials. If you want to read a history of the TEM, the book by Marton (1968) is a compact, personal monograph and that edited by Hawkes (1985) contains a series of individual reminiscences. Fujita (1986) emphasizes the contribution of Japan to the development of the instrument. The field is now at the point where many of the pioneers have put their memoirs down on paper, or Festschrifts have been organized in their honor (e.g., Cosslett 1979, Ruska 1980, and Hashimoto 1986) which detail their contributions over the decades, and compile some useful overview papers of the field. If you enjoy reading about the history of science, we strongly recommend the review of *Fifty Years of Electron Diffraction,* edited by Goodman (1981), and *Fifty Years of X-ray Diffraction,* edited by Ewald (1962). (The spelling of X-ray is discussed in the *CBE Manual,* 1994.)

Figure 1.1. The electron microscope built by Ruska and Knoll in Berlin in the early 1930s.

1.1.B. Microscopy and the Concept of Resolution

When asked what a "microscope" is, most people would answer that it is an instrument for magnifying things too small to see with the naked eye, and most likely they would be referring to the visible-light microscope. Because of the general familiarity with the concept of the light microscope, we will draw analogies between electron and visible-light microscopes wherever it's instructive.

The smallest distance between two points that we can resolve with our eyes is about 0.1–0.2 mm, depending on how good our eyes are, and assuming that there's sufficient illumination to see by. This distance is the *resolution* or *resolving power* of our eyes. So any instrument that can show us pictures (or "images" as we'll refer to them) revealing detail finer than 0.1 mm could be described as a microscope, and its highest useful magnification is governed by its resolution. A major attraction to the early developers of the TEM was that, since electrons are smaller than atoms, it would be possible, at least theoretically, to build a microscope that could "see" detail well below the atomic level. The idea of being able to "see" with electrons may be confusing to you. Our eyes are not sensitive to electrons. If a beam of high-energy electrons was aimed into your eye, you would most likely be blinded as the electrons killed the retinal cells, but you wouldn't see anything! So an integral part of any electron microscope is a viewing screen of some form, which translates electron intensity to light intensity, and which we observe or record photographically. We'll discuss these screens and other ways of recording electron images in Chapter 7.

The resolution of a TEM means different things for different functions of the instrument, and we'll discuss them in the appropriate chapters. It's easiest to think of the image resolution in TEM in terms of the classical Rayleigh criterion for light microscopy, which states that the smallest distance that can be resolved, δ, is given approximately by

$$\delta = \frac{0.61\lambda}{\mu \sin \beta} \qquad [1.1]$$

In equation 1.1, λ is the wavelength of the radiation, μ the refractive index of the viewing medium, and β is the semi-angle of collection of the magnifying lens. For the sake of simplicity we can approximate $\mu \sin \beta$ (which is sometimes called the numerical aperture) to unity and so the resolution is equal to about half the wavelength of light. For green light in the middle of the visible spectrum, λ is about 550 nm (5500 Å), and so the resolution of a good light microscope is about 300 nm. In TEMs we can approximate the resolution in equation 1.1 to $0.61\lambda/\beta$ which, as we'll see later, is very small.

> We'll try to use nanometers throughout this book, but you'll find that many microscopists still insist on using Ångstroms rather than the SI units. However, the Ångstrom is close to the atomic diameter and so is a more convenient unit because it saves us using convoluted phrases like "three tenths of a nanometer."

Now although 300 nm is a small dimension to us it corresponds to about 1000 atom diameters, and therefore many of the features that control the properties of materials are on a scale well below the resolution of the light microscope. So there's a real need to image detail down to the atomic level if we want to understand the properties of materials, and that's a major reason why TEMs are so useful.

This limit of light microscopy was well understood at the turn of this century and prompted Ernst Abbe, one of the giants in the field, to complain that "it is poor comfort to hope that human ingenuity will find ways and means of overcoming this limit." (He was right to be so depressed because he died in 1905, some 20 years before de Broglie's ingenuity solved the problem.) Now de Broglie's famous equation shows that the wavelength of electrons is related to their energy, E, and if we ignore relativistic effects we can show approximately (and exactly in Section 1.4 below) that

$$\lambda \sim \frac{1.22}{E^{1/2}} \qquad [1.2]$$

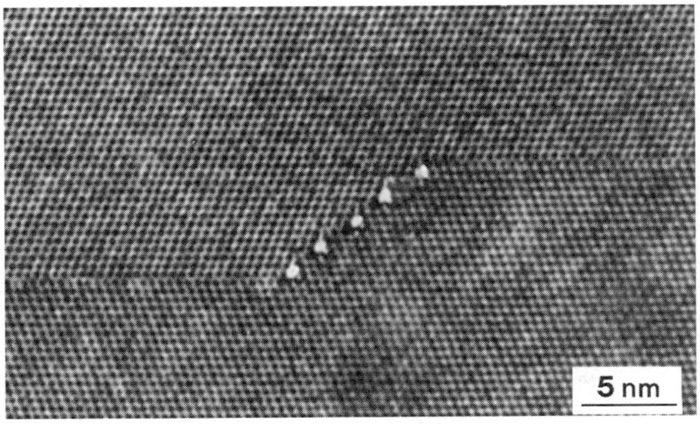

Figure 1.2. A twin boundary in spinel stepping from one {111} plane to another parallel plane. The white dots are columns of atoms. The change in atomic orientation across the twin boundary can be readily seen, even if we do not know what causes the white dots or why, indeed, they are white.

In this equation E is in electron volts (eV) and λ in nm. Remember that we should be precise in our use of the units V and eV: the former represents the *accelerating voltage* of the microscope while the latter refers to the *energy* of the electrons in the microscope. So for a 100-keV electron, we find that $\lambda \sim 4$ pm (0.004 nm), which is much smaller than the diameter of an atom.

We'll see later that we are nowhere near building TEMs that approach this wavelength limit of resolution, because we can't make perfect electron lenses (see Chapter 6). But progress was rapid after Ruska's early work on lenses and, since the mid-1970s, many commercial TEMs have been capable of resolving individual columns of atoms in crystals, creating the field of "high-resolution transmission electron microscopy," or HRTEM, which we'll discuss in Chapter 28. A typical HRTEM image is shown in Figure 1.2. The advantages of shorter wavelengths led in the 1960s to the development of high voltage electron microscopes (HVEMs), with accelerating potentials between 1 MV and 3 MV. In fact, most of these instruments were used to introduce controlled amounts of radiation damage into specimens in an attempt to simulate nuclear reactor environments, but changes in the emphasis of energy research mean there is not much call for such instruments today. While we can still improve the resolution by incremental amounts, the drive for much better resolution is now no longer paramount and the TEM is developing in other ways. In fact, only one HVEM (1 MV) for HRTEM imaging was constructed in the 1980s and three 1.25-MV machines in the 1990s. Intermediate voltage electron microscopes (IVEMs) were introduced in the 1980s. These TEMs operate at 300 or 400 kV, but still offer very high resolution, close to that achieved at 1 MV.

1.1.C. Interaction of Electrons with Matter

Electrons are one type of "ionizing radiation," which is the general term given to radiation that is capable of removing one of the tightly bound inner-shell electrons from the attractive field of the nucleus.

One of the advantages to using ionizing radiation is that it produces a wide range of secondary signals from the specimen, and some of these are summarized in Figure 1.3. Many of these signals are used in "analytical electron microscopy," or AEM, giving us chemical information and a lot of other detail about our samples. AEM uses X-ray energy dispersive spectrometry (XEDS) and electron energy-loss spectrometry (EELS). For example, Figure 1.4A is an X-ray spectrum from a very small region of a TEM specimen showing characteristic peaks which identify the elements present. We can transform such spectra into quantitative data describing elemental changes associated with inhomogeneous microstructures as also shown in Figures 1.4B and C. This aspect comprises Part IV of the book. In contrast, microscopes using nonionizing radiation such as visible light usually only generate light (but not much heat, which is good). AEMs generally offer improved performance at intermediate voltages, similar to HRTEMs.

In order to get the best signal out of our specimens we have to put the best signal in, and so the electron source is critical. We are now very accomplished in this respect as you'll see in Chapter 5, so modern TEMs are very good

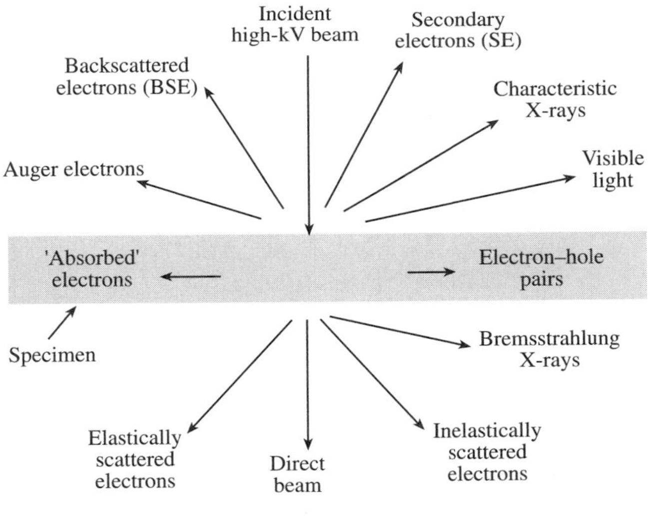

Figure 1.3. Signals generated when a high-energy beam of electrons interacts with a thin specimen. Most of these signals can be detected in different types of TEM. The directions shown for each signal do not always represent the physical direction of the signal but indicate, in a relative manner, where the signal is strongest or where it is detected.

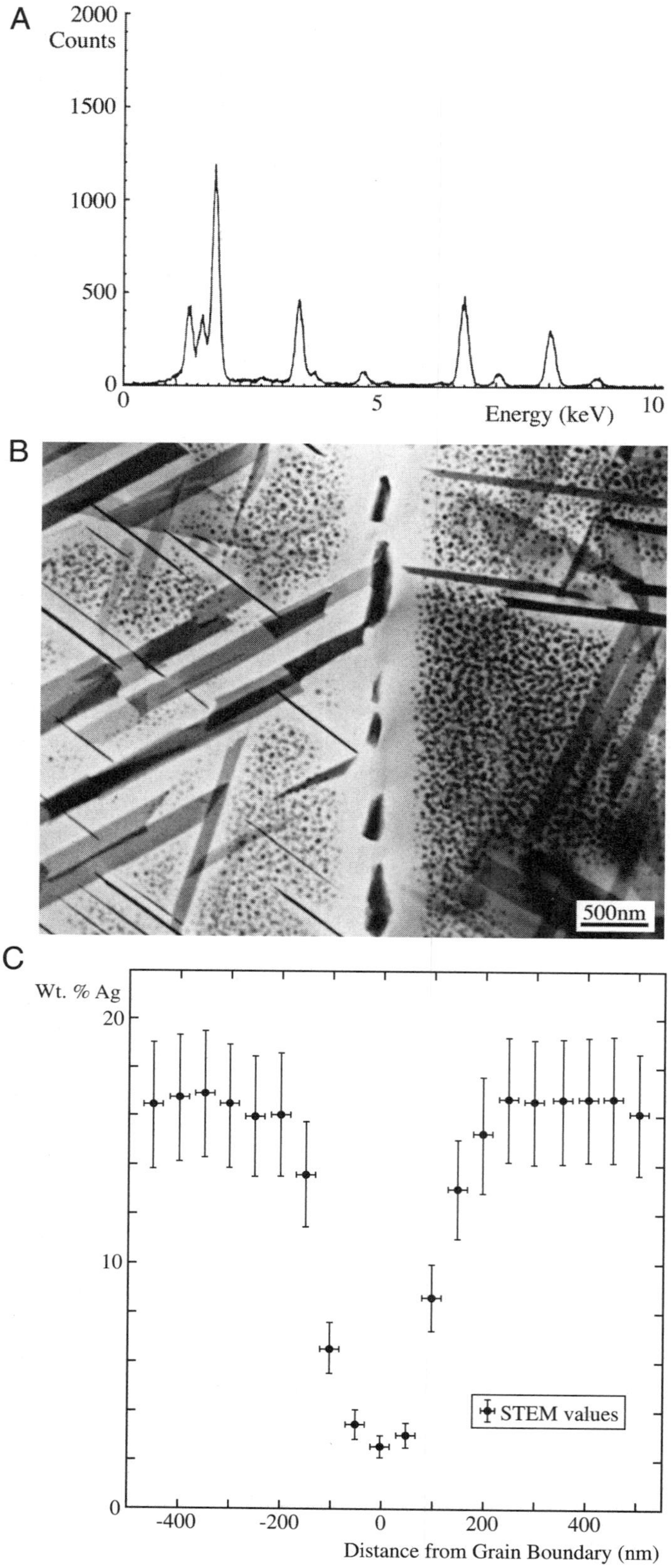

signal-generating instruments. To localize these signals we need to get our TEM to form a very fine electron beam, typically <10 nm and at best <1 nm in diameter. We accomplish this by combining TEM and scanning electron microscope (SEM) technology to create a scanning transmission electron microscope (STEM). The STEM is both the basis for AEMs and a unique scanning imaging microscope in its own right. In fact there are instruments that are only capable of operating in scanning mode and these are sometimes referred to as "dedicated STEMs," or DSTEMs.

1.1.D. Depth of Field

The depth of field of a microscope is a measure of how much of the *object* we are looking at remains "in focus" at the same time. Like the resolution, this property is governed by the lenses in the microscope. The best electron lens is not a very good one, as we've already mentioned, and has been compared to using the bottom of a Coca-Cola bottle as a lens for light microscopy. To minimize this problem we have to use very small limiting apertures in the lenses, narrowing the beam down to a thin "pencil" of electrons which at most is a few micrometers across. These apertures cut down the intensity of the electron beam, but also act to increase the depth of focus of the images that we produce. Remember that "depth of field" refers to the specimen while "depth of focus" refers to the image.

While this large depth of field is chiefly used in the SEM to produce 3D-like images of the surfaces of specimens with large changes in topography, it is also critical in the TEM. It turns out that in the TEM, all of the specimen is usually in focus at the same time, independent of the specimen topography, as long as it's electron transparent! Figure 1.5 shows a TEM image of some dislocations in a crystal. The dislocations appear to start and finish in the specimen, but in fact they are threading their way through the specimen from the top to the bottom, and they remain in sharp focus at all times. Furthermore, we can record the final image at different positions below the final lens of the instrument and it will still be in focus. Compare this with

Figure 1.4. (A) An X-ray spectrum from a small biotite crystal showing peaks at energies that are characteristic of the elements present in the region that interacts with the electron beam. The major peaks from left to right are for Mg, Al, Si, K, Fe, and the Cu support grid. (B) A TEM image of a precipitate-free zone (PFZ) in an aged Al–16 wt% Ag alloy. (C) The Ag profile across the PFZ in (B), obtained through X-ray spectrometry in the TEM showing the depletion of Ag responsible for the PFZ formation.

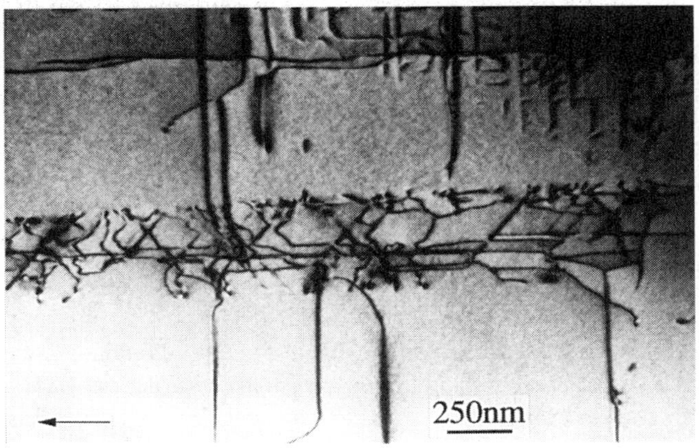

Figure 1.5. TEM image of dislocations in GaAs. A band of dislocations threads through the thin specimen from the top to the bottom but remains in focus through the foil thickness.

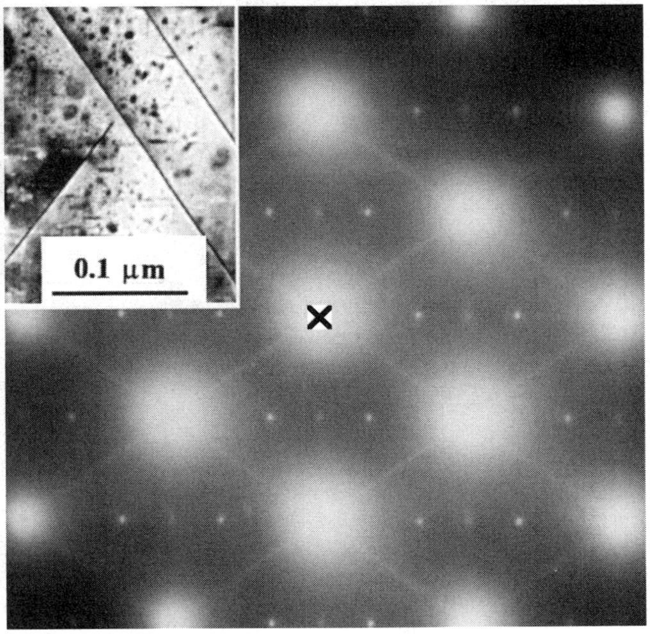

Figure 1.6. TEM diffraction pattern from a thin foil of Al-Li-Cu containing various precipitate phases, shown in the inset image. The central spot (X) contains electrons that come directly through the foil and the other spots and lines are diffracted electrons which are scattered from different crystal planes.

the visible-light microscope where, as you probably know, unless the surface of the specimen is flat to within the wavelength of light, it is not all in focus at the same time. This aspect of TEM gives us both advantages and disadvantages in comparison to the visible-light microscope.

1.1.E. Diffraction

Thompson and Reid showed that electrons could be diffracted when passing through thin crystals of nickel, and the possibility of combining electron diffraction into TEMs was realized by Kossel and Möllenstedt (1939). Today, electron diffraction is an indispensable part of TEM and is arguably the most useful aspect of TEM for materials scientists. Figure 1.6 shows a TEM diffraction pattern which contains information on the crystal structure, lattice repeat distance, and specimen shape, as well as being a most striking pattern. We'll see that the pattern can always be related to the image of the area of the specimen from which it came, in this case shown in the inset. You will also see in Part II that, in addition to the things we just listed, you can conduct a complete crystallographic symmetry analysis of minuscule crystals, including such esoteric aspects as point-group and space-group determination, and at all times the crystallography can be related to the image of your specimen. There is no similar capability on a light microscope because of the relatively large wavelength of visible light.

So an electron microscope can produce atomic level images, can generate a variety of signals telling you about your sample chemistry and crystallography, and you can always produce images that are in focus. There are many other good reasons why you should use electron microscopes. We hope they will become evident as you read through this book. At the same time there are many reasons why you should *not* always seek to solve your problems with the TEM, and it is most important that you realize what the instrument *cannot* do, as well as knowing its capabilities.

1.2. LIMITATIONS OF THE TEM

1.2.A. Sampling

All the above advantages of the TEM bring accompanying drawbacks. First of all, the price to pay for any high-resolution imaging technique is that you only look at a small part of your specimen at any one time. The higher the resolution, therefore, the worse the sampling abilities of the instrument. Von Heimendahl (1980) reported a calculation by Swann in around 1970 estimating that all TEMs, since

Figure 1.7. Photograph of two rhinos taken so that, in projection, they appear as one two-headed beast. Such projection artifacts in reflected-light images are easily discernible to the human eye but similar artifacts in TEM images are easily mistaken for "real" features.

they first became available commercially, had only examined 0.3 mm^3 of material! Extending that calculation to the present time at best doubles the volume to 0.6 mm^3. So we have an instrument that is a terrible sampling tool. This only serves to emphasize that before you put your specimen in the TEM you must have examined it with techniques that offer poorer resolution but better sampling, such as your eyes, the visible-light microscope, and the scanning electron microscope. In other words, know the forest before you start looking at the leaves on the trees.

1.2.B. Interpreting Transmission Images

Another problem is that the TEM presents us with 2D images of 3D specimens, viewed in transmission. Our eyes and brain routinely understand reflected light images but are ill-equipped to interpret TEM images, and so we must be cautious. Hayes (1980) illustrates this problem well by showing a picture of two rhinos, side by side such that the head of one appears attached to the rear of the other (see Figure 1.7). As Hayes puts it: "when we see this image we laugh" (because we understand its true nature in 3D) "but when we see equivalent (but more misleading) images in the TEM, we publish!" So beware of artifacts, which abound in TEM images.

One aspect of this particular drawback is that, generally, all the TEM information that we talk about in this book (images, diffraction patterns, spectra) is *averaged through the thickness of the specimen*. In other words, a single TEM image has no depth sensitivity, as is apparent from Figure 1.5. So other techniques which are surface-sensitive or depth-sensitive, such as field ion microscopy, scanning-probe microscopy, Auger spectroscopy, Rutherford back-scattering, etc., are necessary complementary techniques if you want a full characterization of your specimen.

1.2.C. Electron Beam Damage and Safety

A side effect of ionizing radiation is that it can damage your specimen, particularly in materials such as polymers and some ceramics. Some aspects of beam damage are exacerbated at higher voltages and, with commercial instruments offering up to 400 kV, beam damage now limits much of what we can do in the TEM, even with refractory metals. Figure 1.8 shows an area of a specimen damaged by high-energy electrons. The combination of high-kV beams with the intense electron sources that are available means that we can destroy almost any specimen, if we are not careful. At the same time comes the danger that should *never* be forgotten, that of exposing yourself to ionizing radiation. Modern TEMs are remarkably well engineered and designed with safety as a primary concern, but *never* forget that you are dealing with a potentially dangerous instrument that generates radiation levels that could kill you. So *never* modify your microscope in any way without consulting the manufacturer and without carrying out routine radiation leak tests. If in doubt, don't do it!

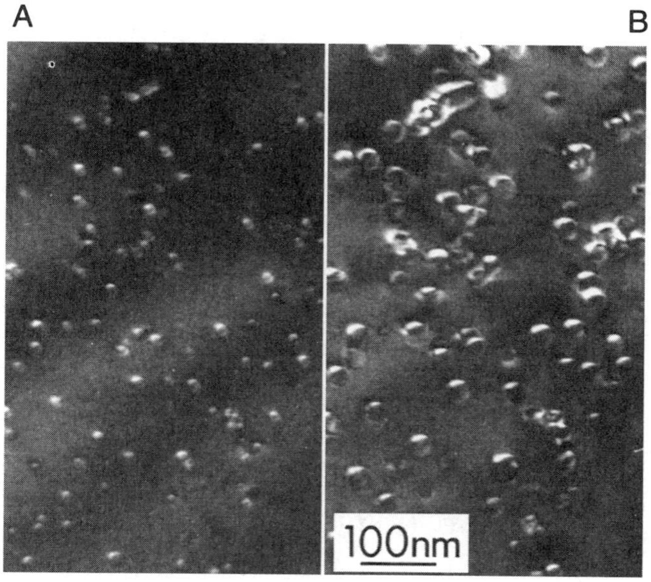

Figure 1.8. Beam damage in quartz after bombardment with 125-keV electrons. With increasing time, from (A) to (B), the damaged regions increase in size.

1.2.D. Specimen Preparation

Your specimens have to be thin if you're going to get any information using transmitted electrons in the TEM. "Thin" is a relative term, but in this context it means "electron transparent." For a specimen to be transparent to electrons it must be thin enough to transmit sufficient electrons such that enough intensity falls on the screen or photographic film to give us an interpretable image in a reasonable time. Generally this requirement is a function of the electron energy and the average atomic number of the specimen. Typically for 100-keV electrons, specimens of aluminum alloys almost up to 1 μm would be thin, while steel would be thin up to about several hundred nm. However, it is an axiom in TEM that thinner is better, and specimens below 100 nm should be used wherever possible, and in extreme cases, such as when doing HRTEM or electron spectrometry, specimen thicknesses <50 nm are essential. These demands become less strict as the beam voltage increases, but this is offset by the danger of beam damage.

> The requirement for thin specimens is a *major* limitation of the TEM. Methods to prepare thin specimens exist for almost all materials, and we talk about them in Chapter 10. But as a general rule the thinning processes that we use do affect the specimen, changing both its structure and its chemistry. So you need to be aware of the dangers of specimen preparation and learn to recognize the artifacts introduced by standard preparation methods.

So it should be obvious to you by now that while TEM and associated techniques are tremendously powerful characterization tools when used properly, they should *never* be used in isolation to solve a materials problem. You must understand your material at low magnification with your eyes and with visible-light microscopy and scanning electron microscopy (SEM) before venturing into TEM studies. Otherwise you may fall foul of some of the limitations we have just listed.

1.3. DIFFERENT KINDS OF TEMs

As you read through the previous sections you will have seen that TEMs come in a wide variety of types: HRTEMs, HVEMs, IVEMs, STEMs, and AEMs. Complete books have been written on each of these instruments, but it is our philosophy that all these are simply different forms of the basic TEM. So in this book we intend to treat them as such. Indeed a current 300 or 400 keV TEM can combine aspects of *all* the above microscope types. Figure 1.9 shows four of the different kinds of TEMs we have mentioned. It is instructive to consider some of the features of the instruments shown here. An HVEM usually requires a two-story room; the scale of each instrument can be judged from the common height of the operator's console. A modern machine essentially is an electron optic column in which we can maintain a good vacuum but the lenses and most other functions can be controlled by one or more computers. Note that the DSTEM only has CRT displays. There is no viewing screen. Furthermore, the electron source is at the base of the column rather than at the top, as we will assume in all of our discussions.

1.4. SOME FUNDAMENTAL PROPERTIES OF ELECTRONS

Many times in the book we'll have to refer to some of the basic properties of electrons. You know that electrons show both particle and wave characteristics, illustrating one of the great puzzles of quantum physics, which we all seem to accept without too much trouble. In fact the TEM routinely demonstrates both the particle and wave characteristics of the electron, repeating the electron analog of G. I. Taylor's famous experiment in which he demonstrated Young's slits interference patterns despite using such a weak light source that only one photon passed through the apparatus at any one time. A typical electron beam current in a TEM is about 0.1–1 μA, which corresponds to about 10^{12} electrons passing through the specimen plane. But as we'll see below, with 100-keV energy, these electrons travel at about $0.5c$ (1.6×10^8 m/s), so they are separated by 0.16 cm and this means that there is never more than one electron in the specimen at any one time. Nevertheless, electron diffraction and interference occur, both of which are wave phenomena, and imply interaction between the different electron beams. Despite this dilemma, we know a lot about the electron and its behavior and some of the basic characteristics are summarized in Table 1.1.

There are a few important equations which you should know. First of all, based on de Broglie's ideas of the wave-particle duality, we can relate the particle momentum p to its wavelength λ through Planck's constant; thus

$$\lambda = \frac{h}{p} \qquad [1.3]$$

In the TEM we impart momentum to the electron by accelerating it through a potential drop, V, giving it a kinetic energy eV. This potential energy must equal the kinetic energy, so

Figure 1.9. Different TEMs: (A) a JEOL 1.25-MV high voltage microscope, used for high-resolution imaging; (B) a Hitachi specialized ultrahigh vacuum TEM for high-resolution surface imaging; (C) a Philips 200-kV analytical microscope with an X-ray spectrometer attached to the stage (the liquid-N_2 dewar cools the detector); and (D) a VG dedicated 100-kV ultrahigh vacuum scanning transmission microscope. Comparison with Ruska's instrument (Figure 1.1) which is 50–60 years older is instructive.

Table 1.1. Fundamental Constants and Definitions

Charge (e)	(−) 1.602×10^{-19} C
1 eV	1.602×10^{-19} J
Rest mass (m_0)	9.109×10^{-31} kg
Rest energy (m_0c^2)	511 keV
Kinetic energy (charge × voltage)	1.602×10^{-19} N m (for 1 volt potential)
Planck's constant (h)	6.626×10^{-34} N m s
1 ampere	1 C/sec
Speed of light in vacuum (c)	2.998×10^{8} m/sec

$$eV = \frac{m_0 v^2}{2} \quad [1.4]$$

Now we can equate the momentum p to the electron mass (m_0) times the velocity (v), and substituting for v from equation 1.4 we obtain

$$p = m_0 v = \left(2m_0 eV\right)^{1/2} \quad [1.5]$$

What all this leads to is the relationship between the electron wavelength λ and the accelerating voltage of the electron microscope, V

$$\lambda = \frac{h}{\left(2m_0 eV\right)^{1/2}} \quad [1.6]$$

This expression is identical to equation 1.2. This relationship between λ and the accelerating voltage introduces a very important concept: by increasing the accelerating voltage we decrease the wavelength of the electrons.

Equations 1.2 and 1.6 are useful expressions for deducing ballpark estimates, but be careful to note the differences. We can use equation 1.6 to calculate the nonrelativistic electron wavelength for typical commercial TEM operating voltages as listed in Table 1.2.

The simple treatment we just went through neglects relativistic effects and, unfortunately for electron microscopists, relativistic effects cannot be ignored at 100-keV energies and above because the velocity of the electrons (as particles) becomes greater than half the speed of light! (The speed of light in vacuum is 2.998×10^8 m/s.) So to be exact we must modify equation 1.6 to give

$$\lambda = \frac{h}{\left[2m_0 eV\left(1 + \frac{eV}{2m_0c^2}\right)\right]^{1/2}} \quad [1.7]$$

A full listing for many more voltages can easily be generated by putting equations 1.6 and 1.7 into a spreadsheet. The effect of relativity is greatest for higher accelerating voltages, as shown in Table 1.2.

There will be many times when it's useful to refer back to these numbers, especially when we consider the resolution of the microscope and when we need to make calculations about the way electrons interact with matter.

A word about units. As we noted above, we should all be using SI units. We don't for two reasons: first, some special units are ideal for the purpose at hand; second, we forget to include special conversion factors in some formulas. The difference between, e.g., the Gaussian system of units and SI units is summarized in the invaluable reference by Fischbeck and Fischbeck (1987).

1.5. MICROSCOPY ON THE INTERNET/WORLD WIDE WEB

TEM users are well integrated into the Internet/WWW and this is a source of useful information (and also some useful knowledge!) about what's going on in the field. Already

Table 1.2. Electron Properties as a Function of Accelerating Voltage

Accelerating voltage (kV)	Nonrelativistic wavelength (nm)	Relativistic wavelength (nm)	Mass ($\times m_0$)	Velocity ($\times 10^8$ m/s)
100	0.00386	0.00370	1.196	1.644
120	0.00352	0.00335	1.235	1.759
200	0.00273	0.00251	1.391	2.086
300	0.00223	0.00197	1.587	2.330
400	0.00193	0.00164	1.783	2.484
1000	0.00122	0.00087	2.957	2.823

you can view research TEMs in real time on the Internet and in due course you'll not only see other instruments but be able to operate them remotely. Such "telepresence microscopy" will represent an extraordinary leap in our ability to characterize materials, since advanced instruments will effectively be available to you in your own laboratories without the need to travel to special sites.

In addition, specialized software packages that allow you to carry out many of the advanced analyses that we will introduce in this text (e.g., diffraction pattern analysis and image/diffraction/spectral simulation) are also available through the Web. In many cases access to this software is limited and any serious microscopy operation should have the software on site, but sometimes it is useful to see the possibilities before you purchase. A list of useful sites is included below but, as with all aspects of the Web, this list is already out of date and the number of actual sites is growing daily.

1.5.A. Microscopy and Microanalysis-Related WWW Sites

http://www.amc.anl.gov

This is the best source for TEM information on the Web in the United States. It is run by N.J. Zaluzec at Argonne National Laboratory (ANL). Through it you can get access to the Microscopy ListServer and a Software Library. There is a connection to the Microscopy & Microanalysis FTP Site and access to Software/Image Libraries. Other useful connections through this site include

http://146.139.72.10/Docs/nonanl/Meetings.html

List of Meetings/Conferences on Microscopy/Microanalysis

http://146.139.72.10/Docs/nonanl/ShortCourses.html

List of Short Courses/Workshops on Microscopy/Microanalysis

http://146.139.72.10/Docs/nonanl/msa/MSA.html

Microscopy Society of America information

http://146.139.72.10/Docs/nonanl/aust/aust.html

Australian Microscopy Societies information

http://146.139.72.10/Docs/nonanl//msc/MSc.html

Microscopical Society of Canada information

http://146.139.72.10/Docs/nonanl/rms/RMS.html

Royal Microscopical Society information

http://146.139.72.10/Docs/nonanl/mas/MAS.html

Microbeam Analysis Society information

http://146.139.72.10/Docs/NonAnl/EduSites.html

University/Educational Sites

http://www.amc.anl.gov/Docs/NonAnl/GovSites.html

Governmental Microscopy Sites

http://146.139.72.10/Docs/NonAnl/ComSites.html

Commercial Sites—microscopy-related manufacturers/suppliers

http://cimewww.epfl.ch/Welcometext.html

A similar operation to the ANL site, but based at the Ecole Polytechnique Fédérale de Lausanne in Switzerland, run by P.-H. Jouneau and P. Stadelmann. A very useful array of software is available including Electron Microscopy Image Simulation (EMS) software, which allows you to perform the following tasks:

> Draw a crystal in perspective view or in projection, stereographic projections, list $(hk\ell)$ distances, structure factors, and extinction distances, draw the microscope contrast transfer function, with $(hk\ell)$ crystal planes, draw kinematical, dynamical, or powder diffraction patterns, draw the amplitude and phase of diffracted beams as a function of specimen thickness, do auto indexing of diffraction patterns, draw Kikuchi patterns, draw high-order Laue zone line patterns, draw convergent beam diffraction patterns (Bloch-wave calculation), draw HRTEM image maps of the crystal (Bloch-wave calculation), do conventional image calculation of dislocations for a cubic crystal

http://cimewww.epfl.ch/emyp/

Another operation based at the Ecole Polytechnique Fédérale de Lausanne in Switzerland, also run by Jouneau and Stadelmann; EM Yellow pages. Contents include:

> Software for Electron Microscopy, Professional Societies, Instruments, Equipment and Consulting, Education in Electron Microscopy, Data and Databases, News and Publications, Related Sources of Information, Conferences, Workshops and Schools, Getting Somewhere Else on the Web

http://www-personal.engin.umich.edu/~jfmjfm/newsgroup.html

Microscopy users newsgroup run by J.F. Mansfield at the University of Michigan

http://www.bocklabs.wisc.edu/imr/microscopists.html

Directory of microscopists on the net

1.5.B. Microscopy and Microanalysis Software

While there is a lot of software available on the WWW, much of it freeware or shareware, you sometimes get what you pay for, so as a serious microscopist you should have access to the best commercial programs, which are not free. Again, this is an aspect of TEM which is changing on a rapid basis, but you can now buy excellent software packages for all the fundamental aspects of microscopy—diffraction, imaging, and microanalysis. Many of these programs will be referenced throughout the text, but here is a brief summary of the best that are currently used, with an indication of the source of the software—some of which are still free! There are many more packages than we have listed here but these are the ones with which we are familiar.

- Comis runs on a UNIX system. A version of the program is being implemented on a Macintosh computer. The graphics are currently supported for Tektronix 4010 and 4100 compatible terminals, but the program may be run from any terminal (without the menu-interface). A version based on the X-window system is also being developed. The interface to the framestore is based on a small set of routines, similar to those used by the SEMPER image processing software (see below). Output to laser printers is supported through the PostScript language. MaComis is available from ComiSoft at 70404.1710@CompuServe.com. Contact the authors for information on the UNIX version
- CRISP is a commercial package running under Windows on a PC. It is designed for image processing of HRTEM images. It can be combined with ELD (see below) and is available from Calidris, Manhemsvägen 4, S-191 46 Solltuna, Sweden (46 8 625 00 41)
- Desktop Microscopist: software for the Macintosh which allows you to calculate diffraction patterns. Available from Virtual Laboratories at http://www. Rt66.com/~virtlabs/ (505 828 1640)
- Differential Hysteresis Imaging: software that enables you to extract the full range of contrast information out of any digitized TEM (or SEM) image. From Klaus Peters at the University of Connecticut, on the Web at http://panda.uchc.edu/htklaus/index.html
- Digital Micrograph: a complete system for the acquisition, control, and processing of digital images from any electron microscope. In principle it can convert an old analog instrument to a digital one, if beam scan coils are available, but it's not worth doing this for a TEM. Use with a CCD camera that provides digital images from the TEM or interface to any STEM system. From Gatan Inc., 6678 Owens Drive, Pleasanton, CA 94588 (510 463 0200)
- DTSA: (Desk-Top Spectrum Analyzer) simulates energy (and wavelength) dispersive X-ray spectra; can also be used as a multichannel analyzer to acquire, interpret, and process spectra. Absolutely essential for the X-ray microanalyst. From NIST, Standard Reference Data Program, 221/A123 Gaithersburg, MD 20899 (301 975 2208)
- ELD is a commercial package from the producers of CRISP running under Windows on a PC. It is intended for quantitative analysis of diffraction patterns and is available from Calidris, Manhemsvägen 4, S-191 46 Solltuna, Sweden (46 8 625 00 41)
- ELP: the energy loss program that runs all Gatan EELS systems. You can't do EELS without it. From Gatan Inc., 6678 Owens Drive, Pleasanton, CA 94588 (510 463 0200)
- EMS: image simulation program, diffraction analysis, basic crystallographic data, and much more developed by P. Stadelmann at Ecole Polytechnique Fédérale de Lausanne and available from him through the Web (http://cimewww.epfl.ch/Welcometext.html). Some would argue that if you have this software, there is little else you need, which is evident from the listing of its capabilities in Section 1.5.A
- Head *et al.* (1973): The book includes a listing of the original source code for the simulation of diffraction contrast images. You can download the source code via the MSA Web page
- Maclispix: a Macintosh-based image processing program, which works in conjunction with NIH-Image (see below). Developed by David Bright at NIST and can be downloaded (free) from the WWW at http://www-sims.nist.gov/WWW/Internet/InternetResources. The FTP command is ftp://enh.nist. gov/mac/mx30a.bin
- MacTempas and CrystalKit: a Macintosh-based image analysis program from Roar Kilaas for the simulation of high-resolution images, diffraction patterns, and crystal structures. Total Resolution, 20 Florida Avenue, Berkeley, CA 94707. roar@totalresolution.com
- Monte Carlo Simulations: software to simulate electron beam trajectories through materials for estimating the spatial resolution of X-ray microanalysis or the backscattered electron yield. Available from David Joy at the University of Tennessee. A full listing is given in his textbook (Joy 1995)
- NCEMSS: (NCEM Simulation System) for HRTEM image simulations, symmetry operators for all 230 space groups, scattering factors for 98 elements. From the National Center for Electron Microscopy, Lawrence Berkeley Laboratories, University of California, Berkeley, CA 94729, or on the Web at http://ncem.lbl.gov/ncem.html
- NIH-Image: public domain software from NIH, developed by Wayne Rasband, for general image manipulation with a limited set of image processing tools. It is useful for grayscale enhancing and Fourier filtering. It can acquire, display, edit, enhance, analyze, print, and animate images. It

reads and writes TIFF, PICT, PICS, and MacPaint files, including programs for scanning, processing, editing, publishing, and analyzing images. It supports many standard image processing functions, including contrast enhancement, density profiling, smoothing, sharpening, edge detection, median filtering, and spatial convolution with user-defined kernels up to 63×63. Available from the Internet by anonymous ftp from zippy.nimh.nih.gov or on floppy disk from NTIS, 5285 Port Royal Rd., Springfield, VA 22161, part number PB93-504868. Details can be found on the WWW at http://rsb.info.nih.gov/nih-image/

- (Adobe) Photoshop and page layout programs for presentations and labeling your figures
- SEMPER (Synoptics, Ltd.): a general-purpose image processing package. It includes many operations suitable for processing of electron microscope images, and has been extended locally. It can be used for Fourier-space filtering of experimental images (including the background-subtraction method for small particles supported on amorphous substrates), for cross-correlation and lattice averaging, and for spatial-frequency enhancement. Image Processing Systems, Synoptics, Ltd., 271 Cambridge Science Park, Milton Road, Cambridge CB4 4WE, UK. Fax: (1223) 420020
- SHRLI: (Simulated High-Resolution Lattice Images) simulations from models up to 3 nm × 3 nm in area (perfect crystal or small defect structures). From the National Center for Electron Microscopy, Lawrence Berkeley Laboratories, University of California, Berkeley, CA 94729, or on the Web at http://ncem.lbl.gov/ncem.html
- Microdiffraction Programs: the listings for several programs are included in Appendix 5 of the book by Spence and Zuo (1992).

CHAPTER SUMMARY

TEMs comprise a range of different instruments which make use of the properties of electrons, both as particles and as waves. The TEM offers a tremendous range of signals from which we can obtain images, diffraction patterns, and several different kinds of spectra from the same small region of the specimen. In the rest of this book we'll take you through the fundamental aspects of electron microscopy, trying to explain at all times *why* we do certain things in certain ways. We'll also explain to some degree *how* we carry out certain operations. Since many different commercial TEMs exist, there's no point in being specific in how to operate the TEM, but we can explain in a generic sense, in many cases, what you have to do to get your microscope to deliver the enormous amounts of information that it generates. Not least of course, we also describe what you need to know to *interpret* the images, diffraction patterns, and spectra that you obtain.

In addition to the WWW, there is a wealth of other sources of information about TEM and, in the general reference list below, we give a selection of appropriate books that emphasize materials science, most of which remain in print, as well as some standard journals and regular conference proceedings.

REFERENCES

General References for TEM

In the reference sections throughout the book, we will list general references that amplify the overall theme of the chapter, as well as specific references that are the source of information referenced in the chapter. If a general reference is referred to specifically in the chapter, we will *not* duplicate it in the specific references.

Books

Amelinckx, S., Gevers, R., and Van Landuyt, J., Eds. (1978) *Diffraction and Imaging Techniques in Material Science,* **1** and **2**, 2nd edition, North-Holland, New York. A collection of excellent individual review articles.

Cowley, J.M., Ed. (1992) *Electron Diffraction Techniques,* **1** and **2**, Oxford University Press, New York. Another collection of excellent individual review articles.

Edington, J.W. (1976) *Practical Electron Microscopy in Materials Science,* Van Nostrand Reinhold, New York. The original out-of-print 1976 edition has been reprinted by TechBooks, 4012 Williamsburg Court, Fairfax, Virginia 22032. It is an essential text, if somewhat outdated.

Goodhew, P.J. and Humphreys, F.J. (1988) *Electron Microscopy and Analysis,* 2nd edition, Taylor and Francis, New York. A succinct summary of SEM, TEM, and AEM.

Hall, C.E. (1953) *Introduction to Electron Microscopy,* McGraw-Hill, New York. A wonderful but nowadays neglected book. The level is very close to this text. Historically minded students will enjoy the Preface.

Hawkes, P.W. and Kasper, E. (1989, 1994) *Principles of Electron Optics,* **1–3**, Academic Press, New York. 1900 pages, comprehensive but advanced. The third volume deals with many aspects of imaging in the TEM, simulation, and processing with ~118 pages of TEM references. An exceptional modern resource.

Heidenreich, R.D. (1964) *Fundamentals of Transmission Electron Microscopy,* Interscience Publisher, New York. Another wonderful but sometimes forgotten classic.

Hirsch, P. B., Howie, A., Nicholson, R.B., Pashley, D.W., and Whelan, M.J. (1977) *Electron Microscopy of Thin Crystals,* 2nd edition, Krieger, Huntington, New York. For many years, the "Bible" for TEM users!

Loretto, M.H. (1994) *Electron Beam Analysis of Materials,* 2nd edition, Chapman and Hall, New York. A concise overview of the subject.

McLaren, A.C. (1991) *Transmission Electron Microscopy of Minerals and Rocks,* Cambridge University Press, New York. Invaluable for the geologist or ceramist.

Reimer, L. (1993) *Transmission Electron Microscopy; Physics of Image Formation and Microanalysis,* 3rd edition, Springer-Verlag, New York. Essential reference text. Strong physics background required.

Sarikaya, M., Ed. (1992) *Resolution in the Microscope, Ultramicroscopy,* **47**. Actually, a collection of reviews which all concern resolution in different instruments.

Sawyer, L.C. and Grubb, D.T. (1987) *Polymer Microscopy,* Chapman and Hall, New York. A broad-based qualitative introduction to TEM and SEM of polymers.

Thomas, G. (1962) *Transmission Electron Microscopy of Metals,* Wiley, New York. A historical volume—the first hands-on book for the materials scientist.

Thomas, G. and Goringe, M.J. (1979) *Transmission Electron Microscopy of Metals,* Wiley, New York. Invaluable for classical imaging and diffraction topics. The original out-of-print 1979 edition has been reprinted by TechBooks, 4012 Williamsburg Court, Fairfax, Virginia 22032.

von Heimendahl, M. (1980) *Electron Microscopy of Materials,* Academic Press, New York. An introductory-level text, no significant AEM or HRTEM component.

Watt, I.M. (1985) *The Principles and Practice of Electron Microscopy,* Cambridge University Press, New York. A basic, practical introduction to SEM and TEM.

Wenk, H.-R. (1976) *Electron Microscopy in Mineralogy,* Springer-Verlag, New York. Required reading for microscopy of geological or ceramic materials.

Williams, D.B. (1987) *Practical Analytical Electron Microscopy in Materials Science,* 2nd edition, Philips Electron Optics Publishing Group, Mahwah, New Jersey. A basic introduction to AEM. The original out-of-print 1987 edition has been reprinted by TechBooks, 4102 Williamsburg Court, Fairfax, Virginia 22032.

Journals

Advances in Imaging and Electron Physics, Academic Press, New York; formerly *Advances in Optical and Electron Microscopy.*

Journal of Microscopy, Blackwell Science, Oxford, United Kingdom.

Microscopy and Microanalysis, Springer, New York (formerly *Journal of the Microscopy Society of America,* Jones and Begell Publishing, Boston.)

Microscopy, Microanalysis, Microstructure (formerly *Journal de Microscopie et Spectroscopie Electronique*), Les Editions de Physique, Les Ulis Cedex A, France.

Microscopy Research and Technique (formerly *Journal of Electron Microscopy Technique*), Wiley-Liss, New York.

Ultramicroscopy, Elsevier Science Publishers, Amsterdam, the Netherlands.

Conference Proceedings

International Congress for Electron Microscopy—every four years (1994).

European Electron Microscopy Congress—every four years (1996).

Microscopy Society of America, San Francisco Press, San Francisco—annual.

Microbeam Analysis Society, VCH, Deerfield Beach, Florida. From 1996, San Francisco Press, San Francisco—annual.

Electron Microscopy and Analysis, Institute of Physics, Bristol, United Kingdom—odd years (1995).

Scanning Electron Microscopy, Scanning Microscopy International, AMF O'Hare, Illinois—annual.

Useful Sources of Numerical Data and Constants

Fischbeck, H.J. and Fischbeck, K.H. (1987) *Formulas, Facts and Constants,* 2nd edition, Springer-Verlag, New York. An invaluable reference. SI units are described in Chapter 2. Relevant equations in Gaussian units are related to SI units on page 127.

Jackson, A.G. (1991) *Handbook for Crystallography for Electron Microscopists and Others,* Springer-Verlag, New York. Ideal for the microscopist, but see the review by A. Eades (*Microsc. Res. Technique* **21**, 368).

Specific References in This Chapter

CBE, Council of Biology Editors (1994) *Scientific Style and Format,* 6th edition, Cambridge University Press, New York.

Cosslett, V.E. (1979) The Cosslett Festschrift, *J. Microsc.* **117**, 1.

Davisson, G. and Germer, L.H. (1927) *Phys. Rev.* **30**, 705.

de Broglie, L. (1925) *Ann. de Physiques* **3**, 22.

Ewald, P.P. (1962) *Fifty Years of X-ray Diffraction,* International Union of Crystallography, Reidel, Dordrecht.

Fujita, H. (1986) *History of Electron Microscopes,* Business Center for Academic Societies, Japan.

Goodman, P. (1981) *Fifty Years of Electron Diffraction,* International Union of Crystallography, Utrecht.

Hashimoto, H. (1986) *J. Electron Microsc. Tech.* **3**, 1.

Hawkes, P.W., Ed. (1985) *The Beginnings of Electron Microscopy, Advances in Electronics and Electron Physics,* Academic Press, New York.

Hayes, T.L. (1980) SEM-1980 **1**, 1 (Ed. O. Johari), SEM Inc., AMF O'Hare, Illinois.

Head, A.K., Humble, P., Clarebrough, L.M., Morton, A.J., and Forwood, C.T. (1973) *Computed Electron Micrographs and Defect Identification,* North-Holland, New York.

Heidenreich, R.D. (1949) *J. Appl. Phys.* **20**, 993.

Joy, D.C. (1995) *Monte Carlo Modeling for Electron Microscopy and Microanalysis,* Oxford University Press, New York.

Knoll, M. and Ruska, E. (1932) *Z. Physik* **78**, 318.

Kossel, W. and Möllenstedt, G. (1939) *Ann. Phys.* **36**, 113.

Marton, L. (1994) *Early History of the Electron Microscope,* 2nd edition, San Francisco Press, San Francisco.

Ruska, E. (1980) *The Early Development of Electron Lenses and Electron Microscopy* (Trans. T. Mulvey), S. Hirzel Verlag, Stuttgart.

Spence, J.C.H. and Zuo, J.M. (1992) *Electron Microdiffraction,* Plenum Press, New York.

Thompson, G.P. and Reid, A. (1927) *Nature* **119**, 890.

Scattering and Diffraction

2

CHAPTER PREVIEW

The electron is a low mass, negatively charged particle. As such, it can easily be deflected by passing close to other electrons or the positive nucleus of an atom. These Coulomb (electrostatic) interactions cause the electron scattering which is the process that makes TEM feasible. We will also discuss how the wave nature of the electron gives rise to diffraction effects. What we can already say is that if the electrons didn't scatter, then there would be no mechanism to create TEM images or diffraction patterns and no source of spectroscopic data. So it is essential to understand both the particle approach and the wave approach to electron scattering in order to be able to interpret all the information that comes from a TEM. Electron scattering from materials is a reasonably complex area of physics, but it isn't necessary to develop a detailed comprehension of scattering theory to be a competent microscopist.

We start by defining some terminology that recurs throughout the book and then we introduce a few fundamental ideas that have to be grasped. These fundamental ideas can be summarized in the answers to three questions.

- What is the probability that an electron will be scattered when it passes near an atom?
- If the electron is scattered, what is the angle through which it is deviated?
- Does the scattering event cause the electron to lose energy or not?

The answer to the first question concerning the probability of scattering is embodied in the ideas of cross sections and mean free paths, so we define these concepts in this chapter. The angle of scattering, usually determined through the differential cross section, is also important because it allows you as TEM operator to control which electrons form the image, and therefore what information is contained in the image. We will develop this point much further when we talk about image contrast in Part III of the book. To answer the third question we must distinguish elastic and inelastic scattering. The former constitutes most of the useful information in diffraction patterns obtained in the TEM, discussed in Part II, while the latter is the source of X-rays and other spectroscopic signals discussed in Part IV. The distinction between electrons that lose energy and those that don't is important enough that we devote the subsequent two chapters to each kind of electron and expand on the basic ideas introduced here.

> The electron beam is treated in two different ways: in electron scattering it is a succession of particles, while in electron diffraction it is treated by wave theory. The analogy to X-rays or visible light would be to compare a beam of photons and an electromagnetic wave. However, we must always remember that electrons are charged particles and that Coulomb forces are very strong.

Scattering and Diffraction

2

2.1. WHY ARE WE INTERESTED IN ELECTRON SCATTERING?

We need to know about electron scattering because it is fundamental to all electron microscopy (not just TEM). You know well that your eye cannot see anything unless it interacts with visible light in some way, for example through reflection or refraction, which are two forms of scattering. Similarly, we cannot see anything in electron microscope images unless the specimen interacts with, and scatters, the electrons in some way. Thus any nonscattering object is invisible, and we will come across situations where "invisibility" is an important criterion. In the TEM we are usually most interested in those electrons that do not deviate far from the incident electron direction. This is because the TEM is constructed to gather these electrons primarily and they also give us the information we seek about the internal structure and chemistry of the specimen. Other forms of scattering, such as electrons which are scattered through large angles, including backscattered and secondary electrons, are also of interest and we will not neglect them, although they are of the greatest interest to SEM users and give surface-sensitive information, such as topography.

In this chapter we introduce the fundamental ideas of electron scattering, then in the next two chapters we discuss the two principal forms of scattering, namely elastic and inelastic. Both forms are useful to us, but you'll see that the latter has the unfortunate side effect of being responsible for specimen damage and ultimately limits what we can do with a TEM.

To give you some feel for the importance of electron scattering, it is worth illustrating at this stage the basic principles of the TEM. You will see in due course that in a TEM we illuminate a thin specimen with electrons in which the electron intensity is uniform over the illuminated area.

> We will often refer to incident and scattered electrons as "beams" of electrons, because we are dealing with many electrons, not an individual electron; these electrons are usually confined to well-defined paths in the microscope. For example, the electron beam that comes through the specimen, parallel to the direction of the incident beam, is an important beam, which we will term the *direct beam.*

As the electrons travel through the specimen, they are either scattered by a variety of processes or they may remain unaffected by the specimen. The end result is that a nonuniform distribution of electrons emerges from the exit surface of the specimen, as shown schematically in Figure 2.1. It is this nonuniform distribution that contains all the structural and chemical information about our specimen. So everything we learn about our specimen using TEM can be attributed to some form of electron scattering.

We'll see in Chapter 9 that the electron microscope is constructed to display this nonuniform distribution of electrons in two different ways. First, the *angular distribution* of scattering can be viewed in the form of scattering patterns, usually called diffraction patterns, and the *spatial distribution* of scattering can be observed as contrast in images of the specimen. A simple (and fundamental) operational step in the TEM is to use a restricting aperture, or an electron detector, of a size such that it only selects electrons that have suffered more or less than a certain angular deviation. Thus, you as operator have the ability to choose which electrons you want to use and thus you control what information will be present in the image. Therefore, to comprehend these images, you have to understand what causes electrons to scatter in the first place.

We devote the whole of Part II to diffraction phenomena and the whole of Part III to images. Then, Part IV deals with ways in which we use inelastic scattering to

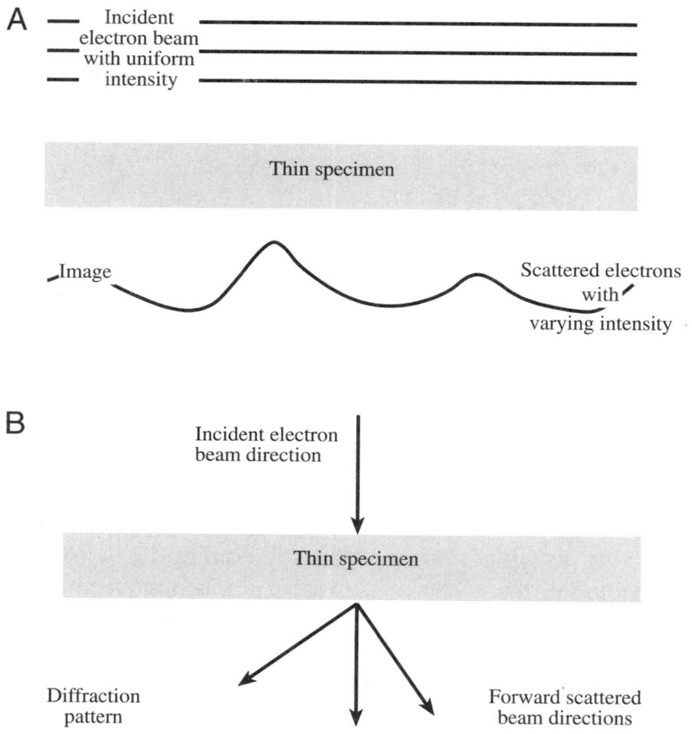

Figure 2.1. (A) A uniform intensity of electrons, represented by the flat line, falls on a thin specimen. Scattering within the specimen changes both the spatial and angular distribution of the emerging electrons. The spatial distribution (intensity) is indicated by the wavy line. (B) The change in angular distribution is shown by an incident beam of electrons being transformed into several forward-scattered beams.

study the chemistry of the specimen. So electron scattering is the theme that permeates this text and connects all aspects of TEM.

2.2. TERMINOLOGY OF SCATTERING

Electron scattering can be grouped in different ways. We've already used the most important terms *elastic* and *inelastic* scattering. These terms are simply descriptions of scattering that results in no loss of energy and some measurable loss of energy, respectively. In this case we tend to consider the electrons as particles, and scattering to involve some interaction like billiard balls colliding. However, we can also separate scattered electrons into *coherent* and *incoherent*, which refers of course to their wave nature. These distinctions are related, since elastic electrons are usually coherent and inelastic electrons are usually incoherent. Let's assume that the incident electron waves are coherent, that is they are essentially in step (in phase) with one another and of a fixed wavelength, governed by the accelerating voltage. (We'll see that this isn't a bad assumption in most circumstances.) Then coherently scattered electrons are those that remain in step and incoherently scattered electrons have no phase relationship after interacting with the specimen.

The nature of the scattering can result in different angular distributions. Scattering can be either *forward scattering* or *back scattering* (usually written as one word) wherein the terms refer to the angle of scattering with respect to the incident beam and a specimen normal to the beam. (Note: you will sometimes see the term forward scattering used in another sense.) If an electron is scattered through < 90° then it is forward scattered, and if > 90° it is backscattered. These various terms are related by the following general principles, summarized in Figure 2.2.

- Elastic scattering is usually coherent, if the specimen is thin and crystalline.

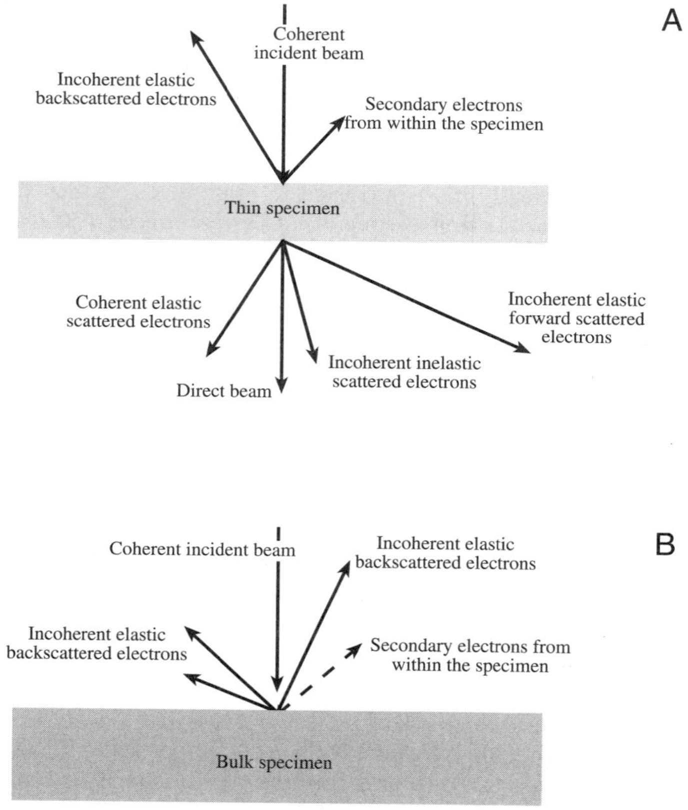

Figure 2.2. Different kinds of electron scattering from (A) a thin specimen and (B) a bulk specimen: a thin specimen permits electrons to be scattered in both the forward and back directions while a bulk specimen only backscatters the incident beam electrons.

- Elastic scattering usually occurs at relatively low angles (1–10°), i.e., in the forward direction.
- At higher angles (> ~10°) elastic scattering becomes more incoherent.
- Inelastic scattering is almost always incoherent and relatively low angle (<1°) forward scattering.
- As the specimen gets thicker, less electrons are forward scattered and more are backscattered until primarily incoherent backscattering is detectable in bulk, nontransparent specimens.

The notion that electrons can be scattered through different angles is related to the fact that electrons can also be scattered more than once. Generally, the more scattering events, the greater the angle of scatter, although sometimes a second scattering event can redirect the electron back into the direct beam so it appears to have undergone no scattering. The simplest scattering process is *single scattering* and we often approximate all scattering within the specimen to a single scattering event (i.e., an electron either undergoes a single scattering event or it suffers no scattering). We'll see that this is often a very reasonable assumption if the specimen is very thin (something you can control). If the electron is scattered more than once we use the term *plural scattering,* and if it is scattered >20 times we say *multiple scattering*. It is generally safe to assume that, unless you have a particularly grim specimen, multiple scattering will not occur. The greater the number of scattering events, the more difficult it is to predict what will happen to the electron and the more difficult it is to interpret the images, diffraction patterns, and spectra that we gather. So once again we emphasize the importance of creating thin specimens so that the single scattering assumption is plausible.

In the transmission electron microscope we utilize the electrons that go through a specimen; it is important to note that such electrons are not simply "transmitted" in the sense of visible light through window glass. Electrons are scattered mainly in the forward direction, i.e., parallel to the incident beam direction. We might ask what percentage of the electrons are forward scattered and how does this vary with the thickness and atomic number of the "target" atom? This scattering is a direct consequence of the fact that there is such a strong interaction between electrons and matter.

> Forward scattering causes most of the signals used in the TEM.

Forward scattering includes elastic scattering, Bragg scattering, the events called diffraction, refraction, and inelastic scattering. Because of forward scattering through our thin specimen, we see a diffraction pattern or an image on the viewing screen, and detect an X-ray spectrum or an electron energy-loss spectrum outside the optical column. But don't neglect backscattering; it is an important imaging mode in the SEM.

2.3. THE CHARACTERISTICS OF ELECTRON SCATTERING

When physicists consider the theory of electron interactions within a solid, they usually consider scattering of electrons by a single, isolated atom, then progress to agglomerations of atoms, first in amorphous solids and then in crystalline solids. When an electron encounters a single, isolated atom it can be scattered in several ways, which we will cover in the next two chapters. For the time being let's imagine simply that, as shown in Figure 2.3, the electron is scattered through an angle θ (radians) into some solid angle Ω measured in steradians (sr). Often we assume that θ is small enough such that sin θ ≈ tan θ ≈ θ. When θ is this small, it is often convenient to use milliradians or mrads; 1 mrad is 0.0573°, 10 mrads is ~0.5°.

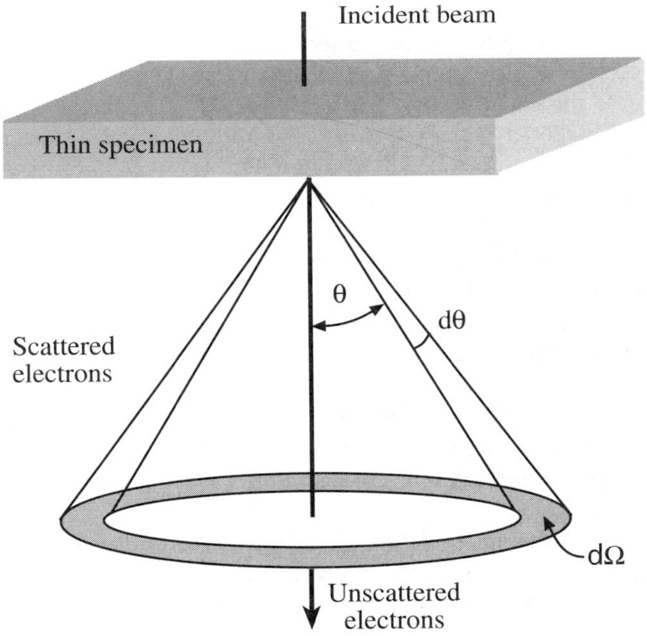

Figure 2.3. Electron scattering by a single isolated atom. The electrons are scattered through a semiangle θ and the total solid angle of scattering is Ω. An incremental increase in scattering angle $d\theta$ gives an incremental increase in a solid angle $d\Omega$.

A convenient definition of a small angle is 10 mrads.

The characteristics of the scattering event are controlled by factors such as the electron energy and the atomic number/weight of the scattering atom. When we consider a specimen rather than a single atom, factors such as the thickness, density, and crystallinity of the specimen also become important. To understand these variables, we need to examine the physics of scattering in more detail.

2.4. THE INTERACTION CROSS SECTION

The chance of a particular electron undergoing any kind of interaction with an atom is determined by an interaction *cross section*. The concept of a cross section is well described by the following analogy given by Rudolf Peierls (Rhodes 1986).

> If I throw a ball at a glass window one square foot in area, there may be one chance in ten that the window will break and nine chances in ten that the ball will just bounce. In the physicist's language this particular window, for a ball thrown in this particular way, has a disintegration (inelastic!) cross section of 0.1 square feet and an elastic cross section of 0.9 square feet.

So each possible interaction has a different cross section which depends on the energy of the particle, in our case the beam energy. The cross section (for which we'll use the letter Q or the Greek σ) has units of area, not square feet as used in Peierls's analogy, but a tiny fraction of a square centimeter termed a "barn." One barn is 10^{-24} cm^2 [that's $(10^{-4}\text{Å})^2$] and the name arises because of the perverse sense of humor of some of the early atomic physicists who considered that this area is "as big as a barn door." Note again the use of non-SI units (cm^2) which persists in the literature, although we can easily define the barn as 10^{-28} m^2. The cross section does *not* represent a physical area, but when divided by the actual area of the atom it represents a *probability* that a scattering event will occur. So the larger the cross section, the better the chances of scattering.

We can look at scattering in two different, but equivalent, ways. First, following Hall (1953), since we are ignoring different kinds of scattering, we can talk about the *total* scattering cross section for the isolated atom, σ_T, which is simply the sum of all elastic and inelastic scattering cross sections such that the total σ_T is

$$\sigma_T = \sigma_{\text{elastic}} + \sigma_{\text{inelastic}} \qquad [2.1]$$

We can define the cross section (an area) in terms of the *effective radius* of the scattering center, r

$$\sigma = \pi r^2 \qquad [2.2]$$

where r has a different value for each of the scattering processes. For example, in the case of elastic scattering, which we'll see is most important in TEM image and diffraction pattern formation, the radius is given as

$$r_{\text{elastic}} = \frac{Ze}{V\theta} \qquad [2.3]$$

where V is the potential of the incoming electron, charge e, which is scattered through an angle greater than θ by atoms of atomic number Z. At first sight this equation seems dimensionally incorrect, since Z and θ are numbers. But V is in volts, so e has to be defined in esu such that the whole term has units of distance. This expression is useful because it indicates the general behavior of electrons in the TEM; i.e., usually electrons scatter less at high kV and high angles, and are scattered more by heavier atoms than light atoms.

Second, following Heidenreich (1964), if we instead consider that the specimen contains N atoms/unit volume, we can define the total cross section for scattering from the specimen (in units of cm^{-1} or m^{-1}) as

$$Q_T = N\,\sigma_T = \frac{N_0\,\sigma_T\,\rho}{A} \qquad [2.4]$$

where N_0 is Avogadro's number (in units of cm^{-3} or m^{-3}), and A is the atomic weight of the atoms in the specimen which has density ρ (so $NA = N_0\rho$). Thus Q can be regarded as the number of scattering events per unit distance that the electron travels through the specimen. If the specimen has thickness t, then the probability of scattering from the specimen is given by

$$Q_T\,t = \frac{N_0\,\sigma_T\,(\rho\,t)}{A} \qquad [2.5]$$

The product of ρ and t is called the "mass-thickness" of the specimen (e.g., doubling ρ produces the same effect as doubling t) and we'll come across this term again when we discuss image contrast and also X-ray absorption. Equation 2.5 is an important expression, since it contains all the variables that affect the scattering probability from a real specimen. We'll use it again when we consider how certain kinds of image contrast arise in the TEM.

Expressions for the cross section become more complicated as they are modified to give better approximations for the scattering in a real specimen. For example, the expression for r_{elastic} which we would substitute in the cross-section equation for elastic scattering (equation 2.2) neglects any screening effects of the electron cloud around the nucleus, which obviously acts to reduce the Z effect. However, the more complex equations don't alter the basic scattering behavior predicted by the simple equations we've just given. If you want to see a fuller description of scattering, then read Chapter 1 by Newbury in Joy *et al.* (1986). If you're a glutton for punishment, the classical text on scattering is by Mott and Massey (1965).

Because of all the variables that affect σ and Q, it is only possible to give a ball-park value for the cross section. For TEM electron energies, the elastic cross section is almost always the dominant component of the total scattering. If you look ahead to Figure 3.3, typical small-angle elastic cross sections for transition metals bombarded by 100-keV electrons are ~10^{-22} m^2 (~10^{-18} cm^2). This is a good number to remember especially when you are considering the probability that a 100-keV electron will be elastically scattered. Inelastic cross sections range from ~10^{-22} m^2 down to 10^{-26} m^2 (100 barns), depending on the specific type of scattering and the material.

2.5. THE MEAN FREE PATH

Instead of using an area to describe the interaction, we can use a length since the distance an electron travels between interactions with atoms is clearly going to be an important concept. The total cross section for scattering can be expressed as the inverse of the mean free path, λ. This new parameter is then the average distance that the electron travels between scattering events. This distance is important because, if we know what it is, we can work out how thin we have to make our specimen so plural scattering is not significant, thus making it easier to interpret our images and spectroscopic data. Because the dimensions of Q are (length)$^{-1}$ there is a simple expression for the mean free path λ which has units of length

$$\lambda = \frac{1}{Q} = \frac{A}{N_0\,\sigma_T\,\rho} \qquad [2.6]$$

Typical values of λ for scattering at TEM voltages are of the order of tens of nm, so single scattering approximations imply specimen thicknesses of this order. It is, unfortunately, conventional to use λ to denote the mean free path; it is *not* the wavelength of the electron. From this equation we can define a probability of scattering p as the electron travels through a specimen thickness t

$$p = \frac{t}{\lambda} = \frac{N_0\,\sigma_T\,(\rho\, t)}{A} \qquad [2.7]$$

which is just $Q_T t$ of equation 2.5.

2.6. THE DIFFERENTIAL CROSS SECTION

Because of the importance of the angle of scattering we need to introduce the concept of the *differential cross section* $d\sigma/d\Omega$. This term describes the angular distribution of scattering from an atom. As shown in Figure 2.3, electrons are scattered through an angle θ into a solid angle Ω and there is a simple geometrical relationship between the θ and Ω

$$\Omega = 2\pi(1 - \cos\theta) \qquad [2.8]$$

and therefore

$$d\Omega = 2\pi \sin\theta \, d\theta \qquad [2.9]$$

So the differential scattering cross section can be written as

$$\frac{d\sigma}{d\Omega} = \frac{1}{2\pi \sin\theta}\frac{d\sigma}{d\theta} \qquad [2.10]$$

Now, we can calculate σ for scattering into all angles which are greater than θ by integrating equation 2.10. This yields

$$\sigma_\theta = \int_\theta^\pi d\sigma = 2\pi\int_\theta^\pi \frac{d\sigma}{d\Omega}\sin\theta\, d\theta \qquad [2.11]$$

The limits of the integration are governed by the fact that the values of θ can vary from 0 to π, depending on the specific type of scattering. If we work out the integral we find that σ decreases as θ increases (which makes physical sense). Since $d\sigma/d\Omega$ is often what we measure experimentally, equation 2.11 gives us an easy way to determine σ for an atom in the specimen: σ for all values of θ is simply the integral from 0 to π. From this we can use equation 2.4 to give us the total scattering cross section from the whole specimen, which we will see later allows us to calculate the TEM image contrast. So we can now appreciate, through a few simple equations, the relationship between the physics of electron scattering and the information we collect in the TEM.

Our knowledge of the values of σ and λ is very sketchy, particularly at the 100–400 keV beam energies used in TEMs. Cross sections and mean free paths for particular scattering events may only be known within a factor of two, but we can often measure θ very precisely in the TEM. We can combine all our knowledge of scattering to predict the electron paths as a beam is scattered through a thin foil.

This process is called Monte Carlo simulation because of the use of random numbers in the computer programs; the outcome is always predicted by statistics!

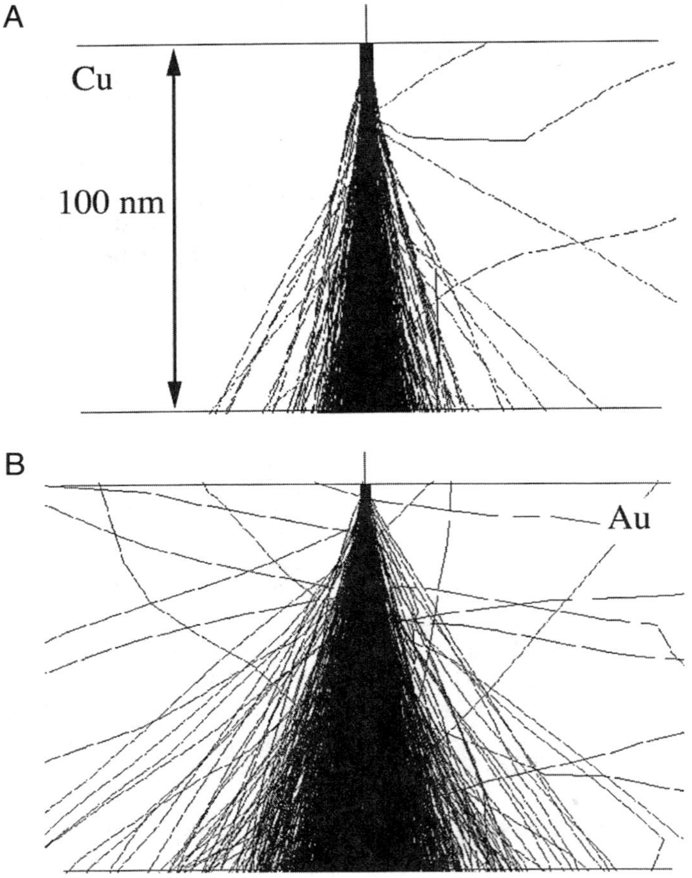

Figure 2.4. Monte Carlo simulation of the paths followed by 10^3 100-keV electrons as they pass through thin foils of (A) copper and (B) gold. Notice the increase in scattering with atomic number.

The Monte Carlo calculation was first developed by two of the United States' foremost mathematicians, J. von Neumann and S. Ulam at Los Alamos in the late 1940s. Ulam actually rolled dice and made hand (!) calculations to determine the paths of neutrons through deuterium and tritium which proved that Teller's design for the "Super" (H-bomb) was not feasible (Rhodes 1995). Monte Carlo methods are more often used in SEM image calculations (see, e.g., Newbury *et al.* 1986, Joy 1995), but they have a role in TEM in determining the expected spatial resolution of microanalysis. Figure 2.4 shows two Monte Carlo simulations of electron paths through thin foils.

2.7. OTHER FACTORS AFFECTING SCATTERING

By selecting electrons of a certain scattering angle (choosing a θ), you are changing the effective scattering cross section (σ_θ), because the scattering strength generally decreases as the angle of scattering increases. Therefore, there will generally be less scattering at higher angles, which explains why we said at the start of the chapter that we are mainly interested in forward scattering in the TEM. Most of the scattered electrons are within ±5° of the unscattered beam.

You also have control of the scattering cross section in other ways. First, the accelerating voltage, which determines the electron energy E_0 (eV), will affect the cross section as implied in equation 2.3 (specifically for elastic scattering). In fact, for all forms of scattering, the total cross section decreases as E_0 increases. Therefore, intermediate and higher voltage TEMs will result in *less* electron scattering than typical 100-kV instruments and, as we'll see in Chapter 4, this has important implications for electron beam damage in delicate specimens, such as polymers.

> Q decreases as E_0 increases. Electron scattering at 300 kV will be smaller than at 100 kV.

We shall see later that the effect of the atomic number of the specimen is more important in elastic than inelastic scattering and, as Z increases, elastic scattering dominates. This behavior helps when we consider ways to enhance scattering (and therefore contrast) in low Z materials such as polymers.

2.8. COMPARISON TO X-RAY DIFFRACTION

There is a very good reason why electrons are used in microscopy: they have a "suitable interaction" with matter. Most descriptions of the interaction of electrons with matter are based on scattering. You will come across such topics as kinematical scattering, dynamical scattering, elastic scattering, inelastic scattering, etc., and we will use the formalism of a scattering factor to describe the process mathematically. It is this scattering process that varies with the structure or composition of the specimen, permitting us ultimately to image a microstructure, record a diffraction pattern, or collect a spectrum. Historically, it was diffraction that provided most of the crystallographic information we have about materials, and the majority of those studies used X-rays. This is why X-ray diffraction is so well documented in the scientific literature. A good understanding of X-ray diffraction helps considerably in understanding electron diffraction; however, the primary processes by which electrons are scattered are very different from the processes by which X-rays are scattered.

X-rays are scattered by the *electrons* in a material through an interaction between the negatively charged electrons and the electromagnetic field of the incoming X-rays. The electrons in the specimen respond to the applied field of the X-ray flux, oscillating with the period of the X-ray beam. These accelerated charged particles then emit their own electromagnetic field, identical in wavelength and phase to the incident X-rays. The resultant field which propagates radially from every scattering source is called the scattered wave.

Electrons are scattered by *both* the electrons and the nuclei in a material; the incoming negatively charged electrons interact with the local electromagnetic fields of the specimen. The incoming electrons are therefore directly scattered by the specimen; it is not a field-to-field exchange as occurs in the case of X-rays. Consequently, electrons are scattered much more strongly than X-rays.

2.9. FRAUNHOFER AND FRESNEL DIFFRACTION

Diffraction of visible light is well understood, so we should carry over as much of the analysis as possible. If you have any experience with diffraction of visible light you will have encountered Fraunhofer and Fresnel diffraction.

- Fraunhofer diffraction occurs when a flat wavefront interacts with an object. Since a wave emitted by a point becomes planar at large distances, this is known as far-field diffraction.
- Fresnel diffraction occurs when it's not Fraunhofer. This case is also known as near-field diffraction.

So why discuss these topics now? We will see later that electron diffraction patterns correspond closely to the Fraunhofer case while we "see" the effects of Fresnel diffraction in our images.

In TEM we will find both forms of diffraction. We will briefly go through the Huygens explanation of how a wave propagates, then consider Fraunhofer diffraction from two slits (Young's slits), and then extend this to many slits. There are two reasons for reviewing this analysis:

- It reminds us that coherent interference is purely a matter of physical optics.
- We can review the concept of phasor diagrams, which we'll use in later chapters.

Huygens explained the propagation of any wavefront by imagining that each point on the wavefront itself acts as a new source for a spherical wavelet. The wavelets interfere with one another to give the new wavefront and the process is repeated.

2.10. DIFFRACTION OF LIGHT FROM SLITS

When we place a pair of very narrow slits in front of a wavefront, we select just two of the Huygens wavelets; these wavelets then must have the same phase at the slits. As they propagate past the slits, their phases differ, depending on the position of the detector. The important term is the path difference $L = d \sin \theta$ as shown in Figure 2.5. The two wavelets propagating in direction $\mathbf{r}$ are out of phase by $2\pi L/\lambda$. If d and λ are such that this phase difference is actually a multiple of 2π, then the rays are again in phase. Therefore, there is an inverse relationship between d and θ for a given d; as d decreases, $\sin \theta$ increases.

> The inverse relationship between *d* and θ occurs solely due to the positions of the slits. We'll come across an identical relationship when we talk about electron diffraction in Section 3.9.

When we extend this analysis to more than two slits we see the same result, but with added subsidiary peaks. The origin of the subsidiary peaks can best be illustrated by considering a series of phasor diagrams. (We'll find similar diagrams useful when we discuss TEM images in Chapter 26.) These diagrams plot the amplitude and phase of the scattered wave as illustrated, for the case of five slits, by the polyhedra in Figure 2.6; in other words, when we add the amplitudes of beams we must take account of their phase. When θ is zero, the rays experience no phase shift and we simply add all of the ampli-

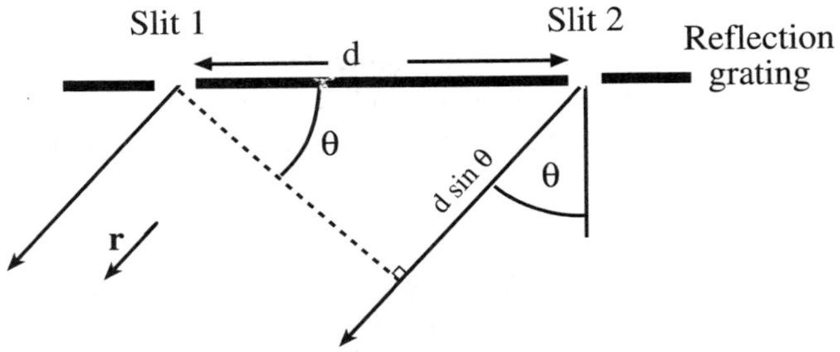

Figure 2.5. An incident plane wave is scattered by two slits, distance d apart. The scattered waves are in phase when the path difference $d \sin \theta$ is $n\lambda$.

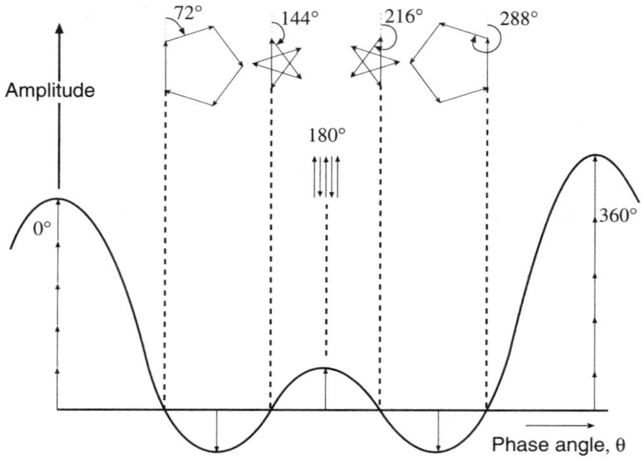

Figure 2.6. A phasor diagram showing how the total amplitude produced by summing five waves produced by five slits varies with the phase angle (how much each wave is out of phase) between the different waves. The arrows show the individual phasors from each slit.

tudes; as θ increases the rays become out of phase, but the phasors can still add to give a large resultant vector. When θ is exactly 72° (360°/5 for five slits), the phasor diagram is a closed pentagon and the resultant amplitude is zero. This process repeats at 144° (2 × 360°/5) and 216° (3 × 360°/5). In between these values at 108° (1.5 × 360°/5) we produce a local maximum in amplitude which is repeated at 180° (2.5 × 360°/5). If we plot the amplitude as a function of θ, we produce the curve with a series of subsidiary maxima shown in Figure 2.6.

Now what happens if we allow the slit to have some width as shown in Figure 2.7A? Each slit produces a phasor diagram as shown in Figures 2.7B and C; i.e., the rays from within a single slit will interfere with each other to modify the polyhedra in Figure 2.6. The amplitude from a single slit varies as $A = \phi^{-1} \sin \phi$, where ϕ is the phase $\lambda^{-1}\pi w \sin \theta$ for a single slit of width w. If we imagine just one slit with width b, we would see a zero in the phasor diagram when $\phi = \pm(n+1/2)\pi$ as shown in Figure 2.7D.

Without going into the detailed math, we can replace the slit of width w by a circular hole or aperture of diameter D. The resulting peak width in the plot of amplitude versus θ then has a maximum of $1.22\lambda D^{-1}$ which is shown in Figure 2.8.

> Because of the circular symmetry of the aperture, the calculation needed to obtain the number 1.22 involves the use of Bessel functions, which you can find in texts on physical optics.

The disk of diameter $1.22\lambda D^{-1}$ is named after Airy and will be one of the fundamental limits on the achievable resolution in TEM, as we discuss in Chapter 6. If we introduce *any* aperture into *any* microscope, we will limit the ultimate resolution of the instrument. As the diameter of the aperture, D, decreases, the minimum resolvable spacing, r, increases. This equation also suggests that decreasing λ (increasing the accelerating voltage) will improve resolution: as λ decreases, r decreases.

The final step is to consider the amplitude scattered from many slits which each have a width, w. The result is shown in Figure 2.9, where we've increased λ/w relative to Figure 2.7.

The important point about this analysis for TEM is that we'll see the same relationship in later chapters, where the slits will be replaced by an aperture, by many atoms, or by a thin specimen.

2.11. COHERENT INTERFERENCE

To expand on this point, consider an infinite plane wave described by the usual characteristics of amplitude and phase. We can describe the wave function for this wave by the expression

$$\psi = \psi_0 e^{i\phi} \qquad [2.12]$$

where ψ_0 is the amplitude and ϕ the phase of the wave. The phase depends on position x, such that if x changes by one wavelength λ, the phase difference is 2π. Stated another way, the phase difference $\Delta\phi$ between any two monochromatic (same wavelength) waves is related to the path difference Δx they must travel in going from source to detector. The relationship is

$$\Delta\phi = \frac{2\pi}{\lambda}\Delta x \qquad [2.13]$$

Figure 2.9 will help you visualize the relationships between the path difference and phase difference for monochromatic waves. Coherent interference between waves relies on the fact that the waves add amplitudes with attention to phase. If all waves scattered by all of the atoms in the specimen are to interfere coherently, they must all differ in phase by integral multiples of 2π. Clearly, this condition requires that the path differences traveled by all of the waves be integer multiples of the wavelength of the incident wave. We can ensure this by requiring that the scattering centers be periodically spaced. Fortunately this is true for all crys-

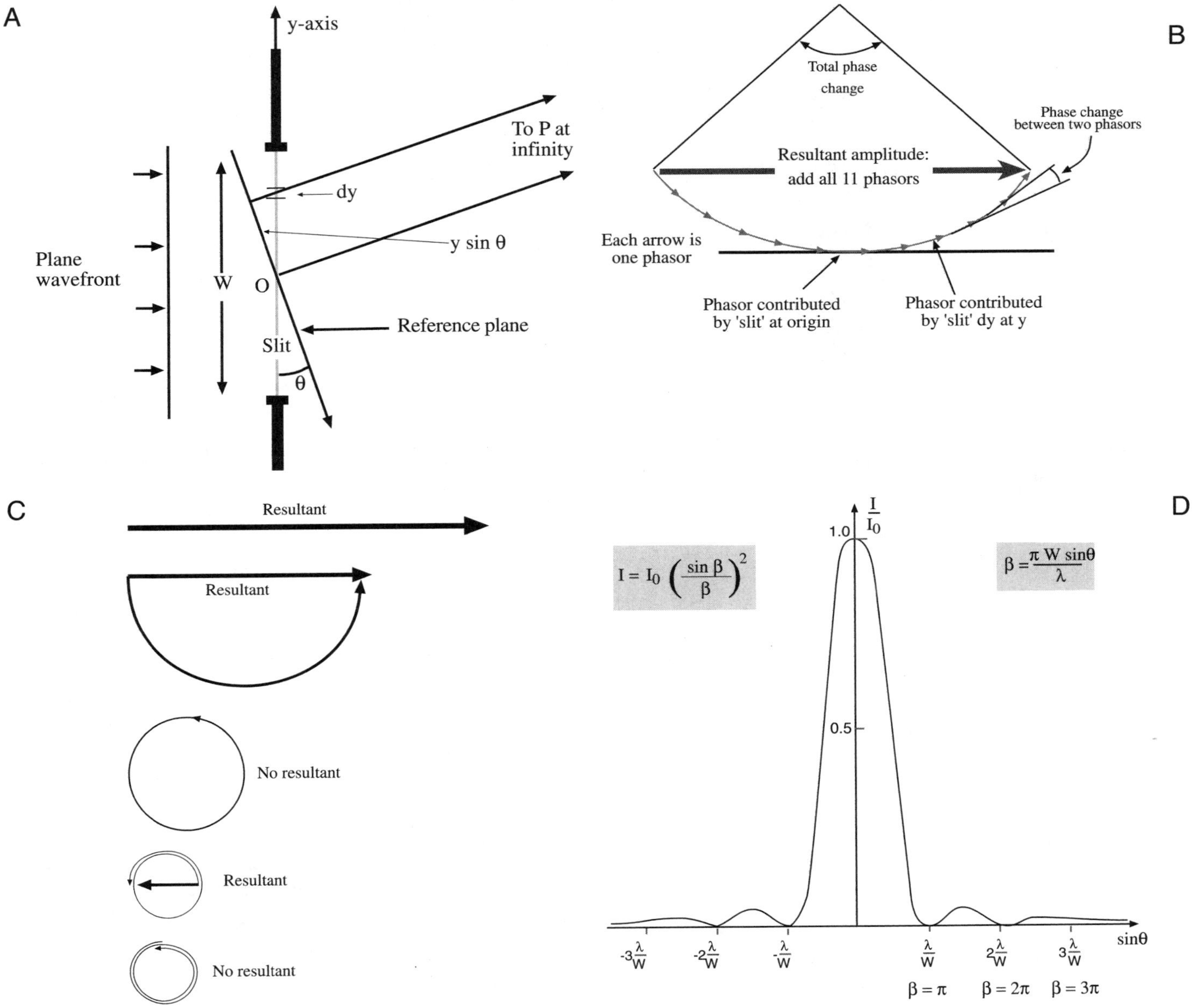

Figure 2.7. (A) Geometry for the scattering from an individual slit. (B) How the phasors from within an individual slit can be added to give the total phasor for the slit shown in (A). (C) How a single slit can produce a beam which has zero amplitude for certain values of θ in (A). The circles are directly comparable to the polyhedra in Figure 2.6. The total length of the phasor increments (from each *dy*) is the same in each figure. (D) A plot of the resulting intensity for scattering from the slit shown in Figure 2.5; this is known as the Fraunhofer diffraction pattern from a single slit; *w* is the slit width defined in (A).

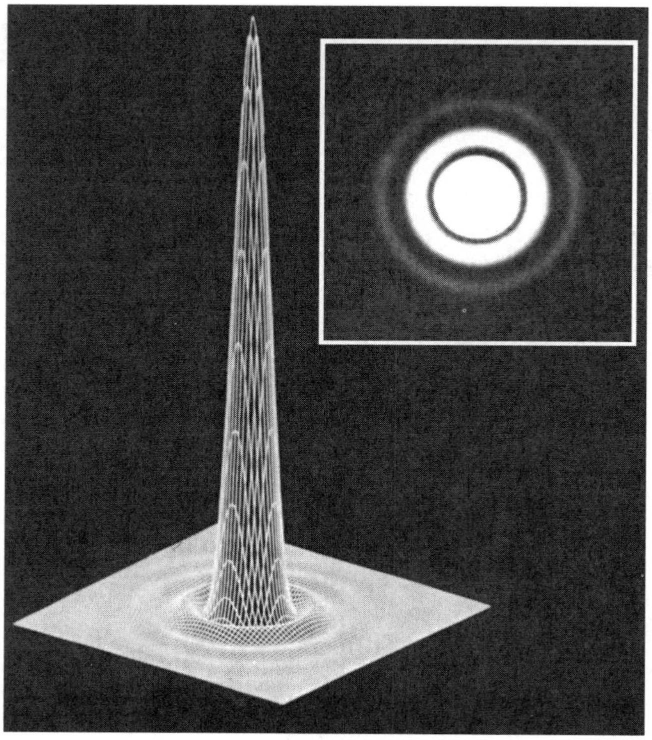

Figure 2.8. The visible-light intensity produced by a 0.5-mm-diameter circular aperture and the observed Airy rings (inset).

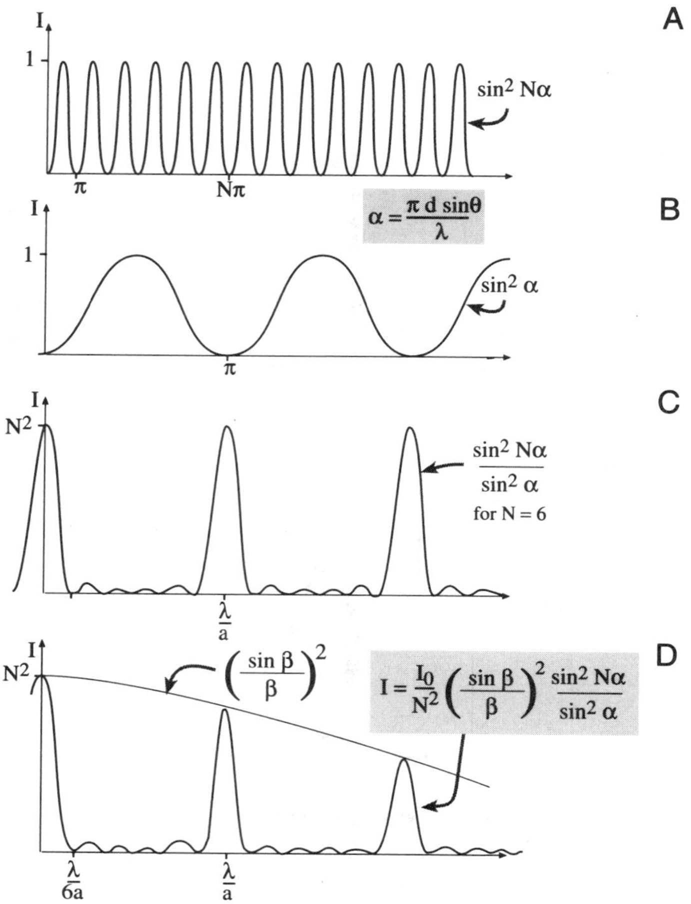

Figure 2.9. The scattered intensity from (A) N slits (shown here for $N = 6$) where each slit would give the intensity shown in (B). (C) is the curve in (A) divided by the curve in (B) and (D) is the curve in (C) multiplied by the curve in (D). The distance d, the separation of the slits, and β are defined in Figures 2.5 and 2.7. (λ/w has been increased compared to Figures 2.7 and 2.8 for simplicity.)

tals, and the mathematical description of coherent interference is simplified (Part II).

2.12. A WORD ABOUT ANGLES

Since angles are so important in the TEM (you can control some of them and the specimen controls others) we want to try to be consistent in our terminology.

- We can control the angle of incidence of electrons on the specimen and we will define the semiangle of incidence as α, as summarized in Figure 2.10.
- In the TEM we use apertures or detectors to collect a certain fraction of the scattered electrons and we will define any semiangle of collection as β.
- We will define all scattering semiangles controlled by the specimen as θ. This may be a specific angle, such as twice the Bragg angle, where $\theta = 2\theta_B$ (see Section 11.4), or a general scattering semiangle θ.

2.13. ELECTRON DIFFRACTION PATTERNS

We've mentioned a couple of times that the TEM is uniquely suited to take advantage of electron scatter because it can form a picture (diffraction pattern) of the distribution of scattered electrons, which we'll discuss in Part II in much more detail. To understand fully how a diffraction pattern is formed in the TEM, you need to go to Chapter 6 to see how electron lenses work, and then to Chapter 9 to find how we combine lenses to form the TEM imaging system. But before we take you through these concepts it is worth just showing a few of the many kinds of diffraction patterns that can be formed in the TEM. At this stage, all

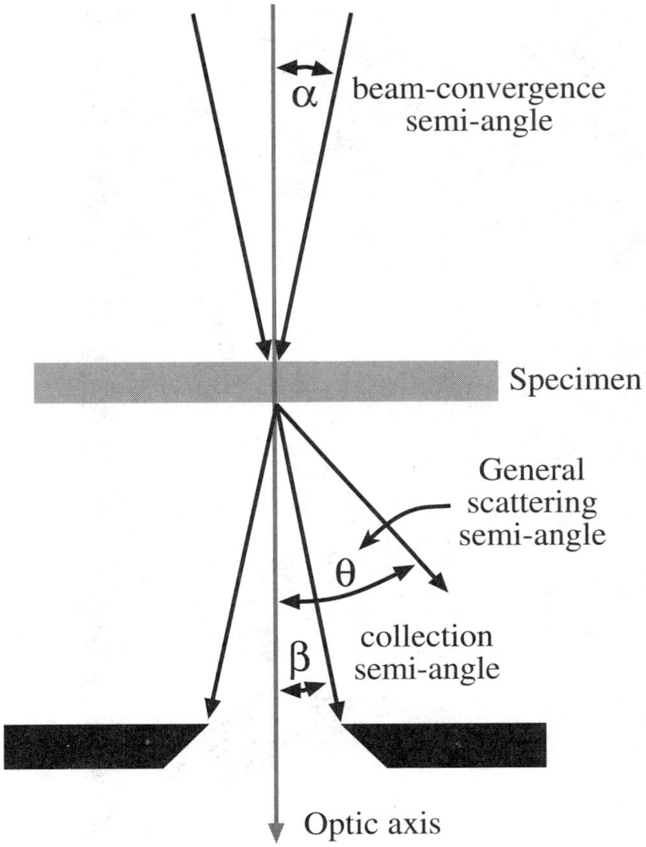

Figure 2.10. Definition of the major semiangles in TEM. Any incidence/convergence semiangle of the beam is termed α; any collection semiangle is β and general scattering semiangles are θ.

you have to do is imagine that a photographic film is placed directly after the thin specimen and that electrons scattered by the specimen as in Figure 2.1B impinge directly on the film. Under these circumstances, the greater the angle of scatter, the further off center the electron hits the film. Thus distances on the film correspond to angles of scatter at the specimen. This relationship is different than the usual interpretation of images in which distances correspond to distances in the specimen, but it is critical to our understanding of diffraction patterns.

Even using this simple description, however, you can comprehend some of the basic features of diffraction patterns. Figure 2.11 is a montage of several kinds of diffraction patterns, all of which are routinely obtainable in a TEM. You can see that several points we've already made about scattering are intuitively obvious in the patterns. First, most of the intensity is in the direct beam, in the center of the pattern, which means that most electrons appear to travel straight through the specimen. Second, the scattered intensity falls with increasing θ (increasing distance from the direct beam), which reflects the decrease in the scattering cross section with θ. Third, the scattering intensity varies strongly with the structure of the specimen. You'll see much more of this in Part II.

So far, in fact, we've only considered the amplitude of the electron wave and we've neglected the phase. When a wave is scattered, it will change its phase with respect to the incident wave. This is because a wave cannot change direction and remain in step with a wave that is unscattered. The phase of the scattered wave is most important in the specific topic of phase-contrast images, which are the principal form of high-resolution atomic-level images such as shown back in Figure 1.2. We'll also come across the importance of the phase of the scattered wave when we consider the intensity of diffracted electron beams and the intensity in diffraction contrast images. But at this stage all you need to know is that the electrons in the beam are in phase when they hit the specimen and the process of scattering, in any form, results in a loss of phase between the scattered and direct beams.

CHAPTER SUMMARY

Remember that electrons are strongly scattered because they are charged particles. This is the big difference compared to X-rays. Thus electrons are scattered by the electron cloud and by the nucleus of an atom. Remember X-rays are only scattered by the electron cloud. (In case you are physics oriented, a quantum mechanical calculation does give the same distribution as the classical calculation for the Coulomb force.)

We have defined three important parameters in this chapter:

σ	the scattering cross section of one atom
Q	the number of scattering events per unit distance traveled in the specimen
$d\sigma/d\Omega$	the differential scattering cross section of one atom

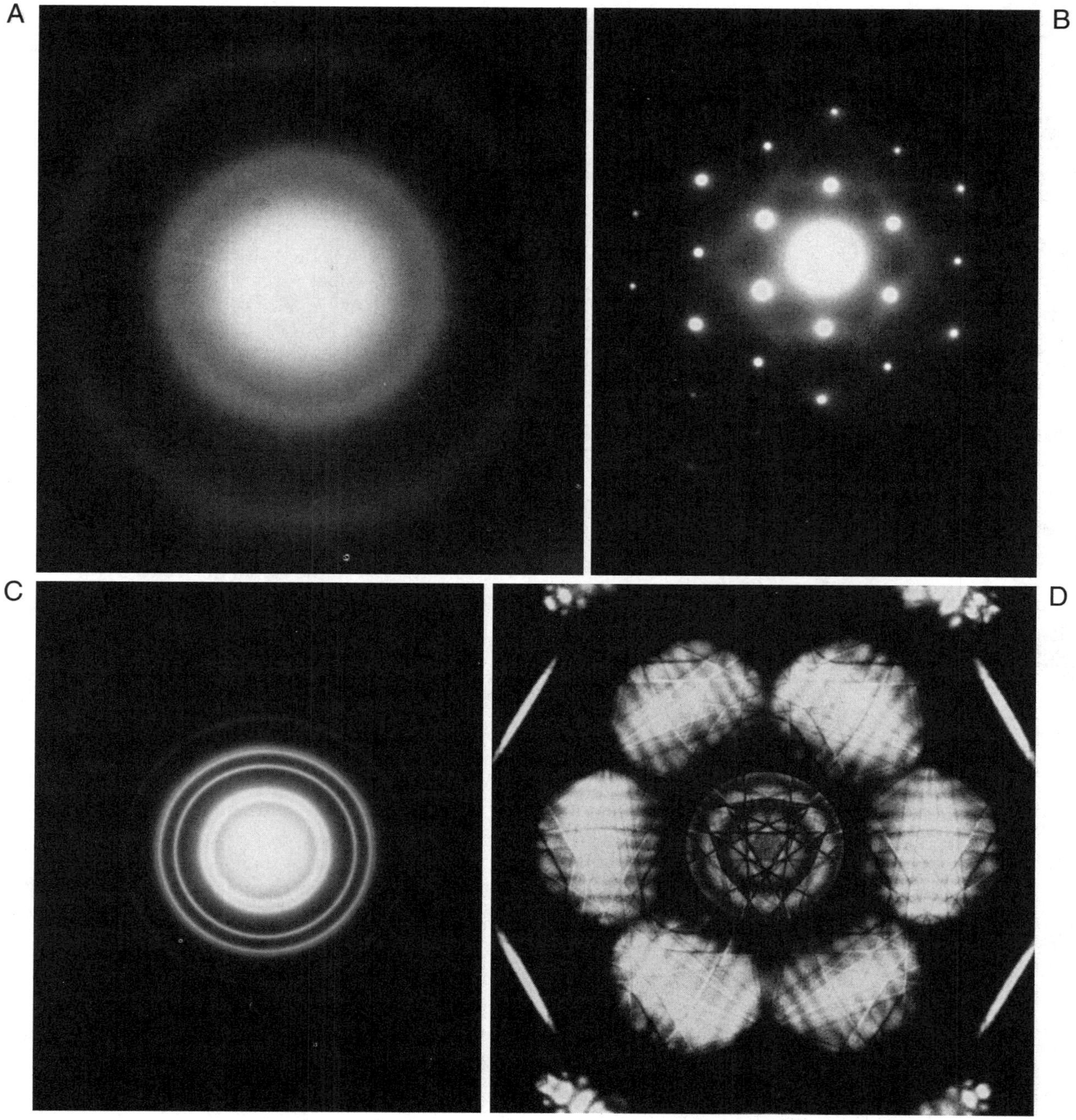

Figure 2.11. Several kinds of diffraction patterns obtained from a range of materials in a conventional 100 kV TEM: (A) amorphous carbon, (B) an Al single crystal, (C) polycrystalline Au, (D) Si illuminated with a convergent beam of electrons. In all cases the direct beam of electrons is responsible for the bright intensity at the center of the pattern and the scattered beams account for the spots or rings that appear around the direct beam.

Finally, a note on grammar! Should we discuss electron scatter or electron scattering? Electrons are scattered and we observe the results of this scattering (a gerund), but in fact we see the scatter (noun) of the electrons which can be measured. However, we'll use electron "scattering" to denote the effect and to be consistent with the popular usage which goes back to the early work of Bragg and others.

REFERENCES

General References

Fishbane, P.M., Gasiorowicz, S., and Thornton, S.T. (1993) *Physics for Scientists and Engineers,* Prentice Hall, Englewood Cliffs, New Jersey.

Goodman, J.W. (1968) *Introduction to Fourier Optics*, McGraw-Hill, New York. Reissued in 1988 and an excellent source for the advanced student.

Hecht, E. (1987) *Optics*, 2nd edition, Addison-Wesley, Reading, Massachusetts.

Klein, M.V. and Furtak, T.E. (1985) *Optics*, 2nd edition, Wiley, New York.

Mott, N.F. and Massey, H.S.W. (1965) *The Theory of Atomic Collisions*, Oxford University Press, New York.

Smith, F.G. and Thomson, J.H. (1988) *Optics*, 2nd edition, Wiley, New York.

Specific References

Hall, C.E. (1953) *Introduction to Electron Microscopy,* McGraw-Hill, New York.

Heidenreich, R.D. (1964) *Fundamentals of Transmission Electron Microscopy,* Interscience, New York.

Joy, D.C. (1995) *Monte Carlo Modeling for Electron Microscopy and Microanalysis*, Oxford University Press, New York

Joy, D.C., Romig, A.D. Jr., and Goldstein, J.I., Eds.(1986) *Principles of Analytical Electron Microscopy*, Plenum Press, New York.

Newbury, D.E., Joy, D.C., Echlin, P., Fiori, C.E., and Goldstein, J.I. (1986) *Advanced Scanning Electron Microscopy and X-ray Microanalysis*, Plenum Press, New York.

Rhodes, R. (1986) *The Making of the Atomic Bomb*, p. 282, Simon and Schuster, New York.

Rhodes, R. (1995) *Dark Sun: The Making of the Hydrogen Bomb*, p. 423, Simon and Schuster, New York.

Elastic Scattering

3

CHAPTER PREVIEW

Elastically scattered electrons are the major source of contrast in TEM images and they also create the intensity distributions in diffraction patterns, so we need to understand what controls this process. We'll consider elastic scattering first from single isolated atoms and then from many atoms together in the specimen. To comprehend elastic scattering we need to invoke both particle and wave aspects of the character of the electron.

Scattering from isolated atoms can occur either as a result of electrons interacting with the negatively charged electron cloud and scattered through small angles of a few degrees, or attracted to the positive nucleus and scattered through large angles up to 180°. The scattering from the nucleus can be interpreted in terms of simple particle–particle collisions, cross sections, and mean free paths that we introduced in the previous chapter. We'll introduce the Rutherford differential cross section, which explains the strong dependence of high-angle elastic scattering on the atomic number (Z) of the atom. Later, we'll use this Z dependence in different ways to form images that reflect the chemistry of the specimen. We can also treat the electrons as waves, in which case their *coherency* becomes important. The coherency of the scattered electrons is related to their *semi-angle* of scattering (θ). As the scattering angle becomes larger, the degree of coherency becomes less and Rutherford-scattered electrons are incoherent.

In contrast to Rutherford high-angle scattering, electrons which are elastically scattered through less than ~3° are coherent. The intensity of this low-angle scattering is strongly affected by the arrangement of atoms within the specimen. Such collective scattering by the atoms is referred to as *diffraction* and can only be understood if we treat the electron as a wave and ignore particle concepts such as cross sections. Diffraction is controlled mainly by the angle of incidence of the electron beam to the atomic planes in the specimen and the spacing of atoms or planes of atoms. So this low-angle scattering is invaluable in characterizing the crystallography of the specimen and is undoubtedly the most significant scattering phenomenon in the TEM.

When we think about the scattering of electrons, we often imagine a beam of particles which hits a target and is deflected to emerge as a beam in another direction, termed the scattered beam, much as we might imagine a beam of light being a group of photons. However, the scattering of light does not always follow the rules of geometric optics because light has a wave character. Similar considerations apply to the diffraction of electrons and this is one of the fundamental concepts of TEM. So you will find the wave-particle duality being used simultaneously, because of both lines of thought.

Elastic Scattering

3

3.1. PARTICLES AND WAVES

We have two different ways of looking at how an electron beam interacts with our specimen in the TEM. We can consider the beam as a succession of particles or as a number of waves. What we want to do is understand the relationship between the two approaches. We can summarize the two viewpoints:

Electrons are *particles* so they have the following properties, which we introduced in Chapter 2.

- They have a scattering cross section and differential scattering cross section.
- They can be scattered through particular angles.
- The electrons interact with the nucleus through Coulomb forces.
- We can relate this process to scattering of other particles, such as α particles, so lots of analysis can carry over from other systems.

> When we discuss *X-ray* and *electron spectrometry* you'll see that we have to use a particle description.

Electrons have a *wave* nature and the electron beam is almost a *plane wave*, hence:

- Waves are diffracted by atoms or "scattering centers."
- How strongly a wave is scattered by an atom is determined by the atomic scattering amplitude.
- We can relate the process to the scattering of X-rays, so lots of analysis already exists.

> When we discuss *imaging, HRTEM,* and *diffraction patterns* you'll see that we use a wave description.

The terminology is sometimes confusing if you look at it closely. A clear definition of diffraction is given by Taylor (1987):

> *An interaction between a wave of any kind and an object of any kind.*

Collins dictionary defines *diffraction* as "a deviation in the direction of a wave at the edge of an obstacle in its path" while *scattering* is defined as "the process in which particles, atoms, etc., are deflected as a result of collision." The word scatter can also be a noun denoting the act of scattering. So scattering might best apply to particles and diffraction to waves; both terms thus apply to electrons! You should also note that the term diffraction is not limited to Bragg diffraction; it refers to any interaction involving a wave.

3.2. MECHANISMS OF ELASTIC SCATTERING

In the previous chapter we simply stated that electrons going through a thin specimen are either scattered or not scattered, and either lose energy or don't lose energy. It's now time to describe the ways in which this scattering occurs and in this chapter we'll confine our attention to elastic events.

It's convenient to divide elastic scattering mechanisms into two principal forms: electron scattering from isolated single atoms and collective scattering from many atoms together within the specimen. We'll start in the same way as we did in the previous chapter by looking first at the interaction of a single electron with an isolated atom. In this situation, elastic scattering can occur in one of two ways, both of which involve Coulomb forces. As shown in Figure 3.1, the electron may interact with the electron

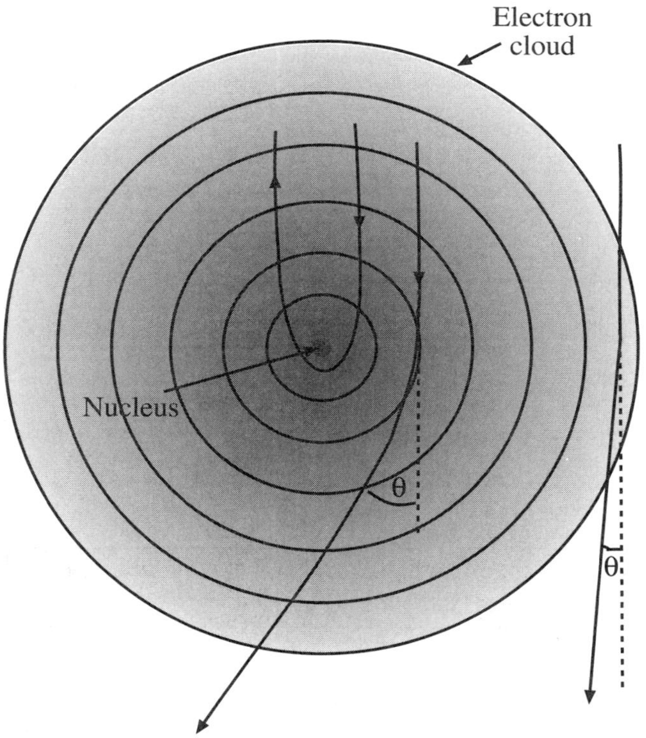

Figure 3.1. Two mechanisms by which a high-energy electron is scattered by an isolated atom. Coulombic interaction within the electron cloud results in low-angle (θ) scatter while Coulombic attraction by the nucleus causes high θ scatter and perhaps complete backscatter. The potential within the electron cloud is always positive.

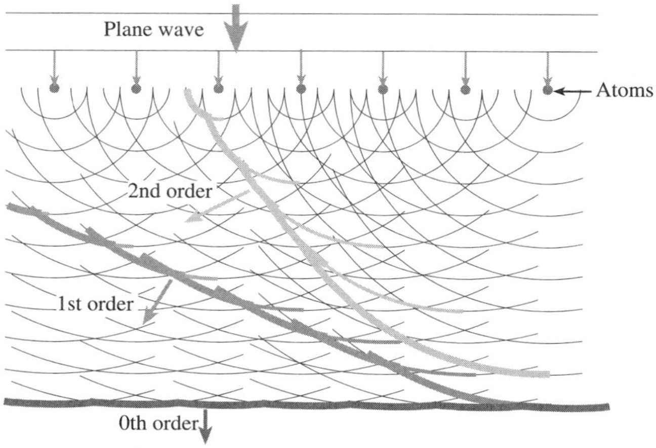

Figure 3.2. A plane coherent electron wave generates secondary wavelets from a row of scattering centers (e.g., atoms in the specimen). The secondary wavelets interfere, resulting in a strong direct (zero-order) beam and several orders of coherent beams scattered (diffracted) at specific angles.

cloud, resulting in a small angular deviation. Alternatively, if an electron penetrates the electron cloud and approaches the nucleus, it will be strongly attracted and may be scattered through a larger angle that in rare cases in the TEM can approach 180° (complete backscattering).

> You should be aware that either of these two interactions may not be truly elastic, so our separation of scattering into elastic and inelastic is a bit of a simplification.

In fact many electron–electron interactions are inelastic, as we'll see in the next chapter. We'll also see, for example, that the nuclear interaction may result in the generation of a bremsstrahlung X-ray, or may even result in the displacement of the atom from its site in the crystal, both of which involve some energy loss for the electron. Indeed, the higher the angle of scattering of an electron emerging from the specimen, the greater the chance that it will have undergone an inelastic event at some time during its passage through the specimen. Despite all this, we'll ignore any inelastic effects in this chapter.

The second principal form of elastic scattering occurs when the electron wave interacts with the specimen as a whole. We've already mentioned the best known form of this interaction, namely diffraction, which is particularly important at low angles. Understanding diffraction involves treating the electron beam as a wave, rather than as a particle as we did in Figure 3.1. Following the original approach of Huygens for the diffraction of visible light, we imagine each atom in the specimen acting as a source of "secondary" spherical wavelets as illustrated in Figure 3.2. These wavelets reinforce one another in certain angular directions and cancel in others. Thus the low-angle elastic scattering distribution is modified by the crystal structure of the specimen, and intense diffracted beams emerge at certain specific angles; we'll discuss these higher-order effects in Chapters 11 and 12. We'll now go on to examine these two forms of elastic scattering in more detail, starting with the simplest concept which is sometimes referred to as the billiard-ball model. We will briefly describe the scattering of a wave to show how it relates to this particle-scattering treatment and use this as the basis later for a full analysis of diffraction.

3.3. SCATTER FROM ISOLATED ATOMS

Consider the two paths for an electron passing close to an atomic nucleus shown in Figure 3.1. In either case, the direction traveled by the electron changes; the electron is scattered through an angle θ.

> Elastic electron–electron interactions usually result in a relatively low scattering angle, while electron–nucleus interactions cause higher-angle scattering.

If we just consider an electron, charge e, scattering from an isolated atom, the electron–electron and electron–nucleus scattering cross sections can be easily expressed by two very simple equations (Hall 1953)

$$\sigma_{\text{electron}} = \pi r_e^2 = \pi\left(\frac{e}{V\theta}\right)^2 \qquad [3.1]$$

$$\sigma_{\text{nucleus}} = \pi r_n^2 = \pi\left(\frac{Ze}{V\theta}\right)^2 \qquad [3.2]$$

Remember that we are using a billiard-ball model where r_e and r_n are the radii of the electron cloud and the nucleus, respectively. Hopefully no one will ask you to prove these simple expressions.

You can see that the atomic number Z of the atom controls the elastic interaction with the nucleus, but the electron–electron scattering is more a function of the incident beam energy (V in volts). We'll see later in Chapter 22 that the strong effect of Z becomes important when we need to enhance scattering in low-Z materials, such as polymers and biological tissue, in order to get better TEM image contrast. Notice that when the electron passes close to the nucleus (r_n is small) the angle θ will be large. We'll see in Chapter 22 that this dependence on θ directly relates to TEM-image contrast. The electron beam energy can also control the image contrast to some extent. For the time being, we'll ignore the low-angle electron cloud scattering and concentrate only on scattering by the nucleus.

3.4. THE RUTHERFORD CROSS SECTION

The high-angle electron–nucleus interaction is analogous to the backscattering of α particles from a thin metal foil. The first observation of such backscattering in 1911 by H. Geiger (of *counter* fame) and a Manchester University *undergraduate*, E. Marsden, enabled their professor, Rutherford, to deduce the existence of the nucleus (never overlook undergraduate research results!). Rutherford (1911), describing backscattering as "the most incredible event that has ever happened to me," derived the following expression for the differential cross section for this kind of scattering

$$\frac{d\sigma(\theta)}{d\Omega} = \frac{e^4 Z^2}{16(E_0)^2 \sin^4\frac{\theta}{2}} \qquad [3.3]$$

All the terms in this equation were defined in Chapter 2. The expression assumes that the incident electron does not lose significant energy through inelastic processes, so that the energy of the electrons, E_0 (in keV), is fixed. This is generally a good assumption in the TEM. We can substitute appropriate values for the various constants and integrate the differential cross section from 0 to π to obtain the total elastic nuclear cross section (in scattering events per electron per atom per m^2) in a more accurate form than that given in equation 3.2

$$\sigma_{\text{nucleus}} = 1.62 \times 10^{-24}\left(\frac{Z}{E_0}\right)^2 \cot^2\frac{\theta}{2} \qquad [3.4]$$

Again we see that the beam energy (E_0), the angle of scattering (θ), and the atomic number (Z) all affect the probability that an electron will be scattered by the nucleus. As in Chapter 2, we can modify this expression for a single isolated atom to take into account the scattering from atoms in a TEM specimen of thickness t

$$\begin{aligned} Q_{\text{nucleus}}\, t &= \left(N_0\frac{\rho}{A}t\right)\sigma \\ &= 1.62\, x\, 10^{-24}\,\left(N_0\frac{\rho}{A}t\right)\left(\frac{Z}{E_0}\right)^2\cot^2\left(\frac{\theta}{2}\right) \end{aligned} \qquad [3.5]$$

Notice that we still have the mass-thickness dependence, ρt, but that the strong dependence on Z is now obvious (compare to equation 2.5). See Joy *et al.* (1986) for further discussion of these calculations.

3.5. MODIFICATIONS TO THE RUTHERFORD CROSS SECTION

You'll often see the Rutherford differential cross section (equation 3.3) in different, but mathematically similar, forms. The expression given here neglects the screening effect of the surrounding electron cloud, which acts to reduce the differential cross section, thus lowering the amount of scattering. In other words, when the electron does not pass close to the nucleus, the scattering angle will be small (say < 5°) because screening is important. If we wish to account for screening, we replace the $\sin^2(\theta/2)$ term with $[\sin^2(\theta/2) + (\theta_0/2)^2]$, where θ_0 is called the screening parameter given by

$$\theta_0 = \frac{0.117\, Z^{1/3}}{E_0^{1/2}} \qquad [3.6]$$

(Here E_0 is in keV.) What we are saying is that the screening parameter can be described by a particular scattering angle, θ_0. When the scattering angle is greater than θ_0, we can neglect electron–electron interactions and the nuclear interaction is dominant. The value of θ_0 at 100 keV is only ~ 5° for Cu and less for lighter elements, so above this angle all scattering can be approximated to Rutherford-type high-angle scattering.

You should note that equation 3.3 is nonrelativistic, which is unfortunate since relativistic effects are significant for electrons with energies greater than 100 keV, as in the TEM. However, we can correct for relativity to give a more accurate differential cross section which is expressed using λ_R, the relativistically corrected wavelength (see equation 1.7), and a_0, the Bohr radius of the scattering atom

$$a_0 = \frac{h^2 \varepsilon_0}{\pi\, m_0 e^2} \qquad [3.7]$$

where ε_0 is the dielectric constant. Using the other constants listed in Table 1.1 we find a_0 is 0.0529 nm (remember this as 0.5 Å). The result is

$$\frac{d\sigma(\theta)}{d\Omega} = \frac{\lambda_R^{\,4}\, Z^2}{64\pi^4 (a_0)^2 \left(\sin^2\frac{\theta}{2} + \left(\frac{\theta_0}{2}\right)^2\right)^2} \qquad [3.8]$$

> This is called the screened relativistic Rutherford cross section.

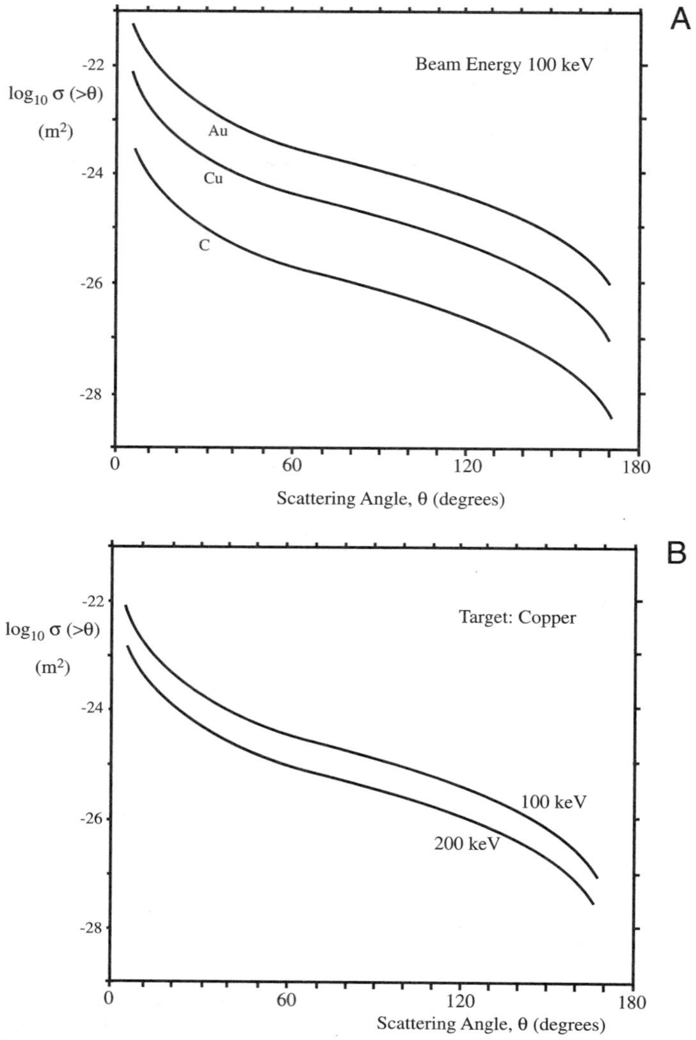

Figure 3.3. The variation of the logarithm of the screened relativistic Rutherford cross section with scattering angle: (A) for different elements at 100 keV and (B) for Cu at different TEM voltages.

This cross section is the one most widely used for TEM calculations, although it has particular limitations at the highest TEM voltages (300–400 kV) and for the heavier elements (Z>30), which cause large scattering angles. Under these circumstances you should use another cross section, such as that of Mott, for which you should consult the text by Mott and Massey (Chapter 2).

The best way to summarize the characteristics of cross sections is to present some data. Figure 3.3 shows the variation of the screened Rutherford cross section with scattering angle for three different elements and two different electron-beam energies. As you can see, the cross section decreases by several orders of magnitude from ~10^6 barns to about 10 barns as the scattering angle increases from 0–180°; so, as expected, scattering is most likely to occur in the forward (θ close to 0°) direction. Doubling the electron-beam energy can lower the cross section by a factor of two or three, which confirms that higher-energy electrons are less likely to be scattered, all else being equal. Figure 3.4 plots the related mean free paths for elastic scattering. From this graph you can see that very few high-angle elastic scattering events will occur if you keep your specimen below 100-nm thickness and you can then approach the ideal of single scattering that we'll assume many times throughout this text.

3.6. COHERENCY OF THE RUTHERFORD-SCATTERED ELECTRONS

Up to now we've treated the electron as a particle, but there is useful insight to be gained if we examine the wave nature of the scattered electron. The coherency of the scattered electron wave is a distinguishing characteristic. High-angle

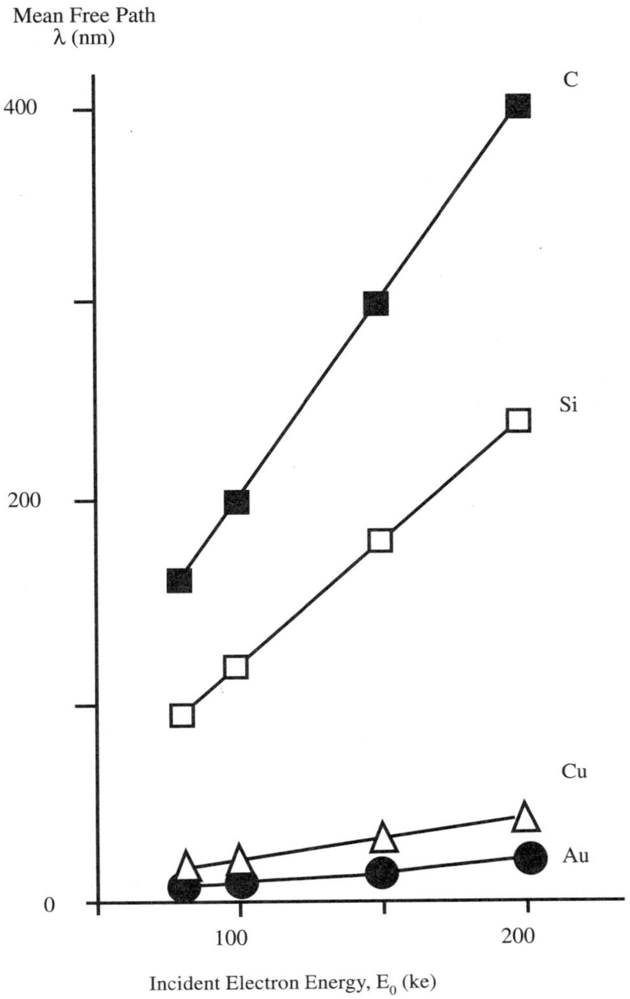

Figure 3.4. The variation of the mean free paths of elastic scatter for four different elements as a function of the beam energy, calculated assuming a screened, relativistic Rutherford cross section.

Rutherford-scattered electrons are *incoherent:* the phases of the electron waves are not in step. Such incoherent scattering is important in two respects. First, the high-angle forward scattering can be used to form exceptionally high-resolution images of a crystalline specimen in which the image contrast is due to the value of Z, not the orientation of the specimen, as we'll see is the case for low-angle coherent scattering. Such *Z-contrast images,* as we'll see in Chapter 22, provide qualitative atomic resolution microanalysis in addition to showing atomic resolution detail at interfaces between regions of different Z. This is a relatively new imaging technique which may revolutionize our understanding of materials. Second, the high-angle backscattered electrons can be used to form images of the beam entrance surface of the specimen in which the contrast is not only due to differences in Z, but also to changes in surface topography of the specimen. Backscattered electron images are rarely used in the TEM because the backscattered signal is small. If you go back and look at the Monte Carlo simulation in Figure 2.4, you'll see that out of 10^3 incident electrons in Cu only about three (0.3%) were backscattered, and only one of these escaped from the foil. Therefore, the quality of the signal is very poor and the images are noisy. This contrasts with bulk specimens in an SEM in which many electrons are backscattered (about 30% in Cu).

3.7. THE ATOMIC SCATTERING FACTOR

The classical Rutherford differential cross section cannot be used to calculate the cross section exactly, because it ignores the wave nature of the electron beam. A full treatment involves wave mechanics and is well beyond the scope of this text.

Perhaps the most familiar aspect of the wave approach to a cross section is the concept of the atomic scattering factor $f(\theta)$. The atomic scattering factor is related to the differential elastic scattering cross section, thus

$$|f(\theta)|^2 = \frac{d\sigma(\theta)}{d\Omega} \qquad [3.9]$$

What we will do is to highlight some of the important features that lead to this result by outlining the arguments.

- $f(\theta)$ is a measure of the amplitude of an electron wave scattered from an isolated atom.
- $|f(\theta)|^2$ is proportional to the scattered intensity.

From these two statements you can appreciate why $f(\theta)$ is such an important parameter.

The scattering-factor approach is complementary to the Rutherford differential cross section, because it is most useful for describing the low-angle elastic scattering where the Rutherford model is inappropriate. Usually, the atomic scattering factor is defined in the following manner

$$f(\theta) = \frac{\left(1 + \frac{E}{E_0}\right)}{8\pi^2 a_0}\left(\frac{\lambda}{\sin\frac{\theta}{2}}\right)^2 (Z - f_x) \qquad [3.10]$$

where all the terms have been previously defined. If you need a more detailed approach you could consult the text by Reimer (1993). The wavelength is λ and f_x is the scattering factor for X-rays, which is well known. The best source of electron scattering factors is that due to Doyle and Turner

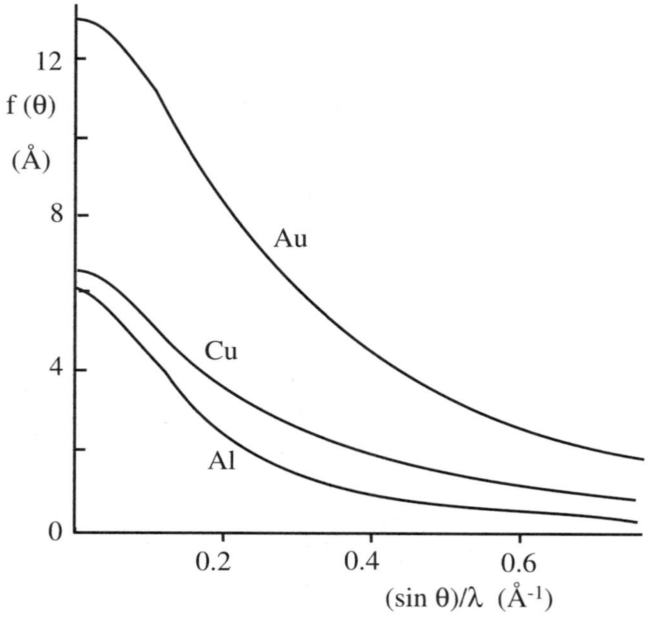

Figure 3.5. Change in the atomic scattering factor $f(\theta)$ with scattering angle θ (calculated from equation 3.10) showing that elastic scattering decreases with angle away from the incident beam direction (θ = 0°) and increases with *Z*.

(1968), and you can also find values in the NCEMSS software (Section 1.5). The appearance of f_x in this formula is a reminder that $f(\theta)$ is a fundamental result of the wave nature of the electron.

> $f(\theta)$ depends on λ, θ, and *Z*.

We can plot this angular variation for a single isolated atom. Figure 3.5 summarizes graphically what we already know about elastic scattering:

- It decreases as θ increases (θ = 0° for the incident beam direction).
- It decreases as λ decreases (i.e., as the accelerating voltage increases).
- It increases with *Z* for any value of θ.

> The important point to remember is that both the differential cross section and the scattering factor are simply a measure of how the electron scattering intensity varies with θ.

This expression (equation 3.10) for $f(\theta)$ contains components of both elastic nuclear scattering (the *Z* term) and elastic electron-cloud scattering (the f_x term). We'll see later in the chapters on diffraction that the $f(\theta)$ approach is used exclusively, and if we neglect the f_x term then it can be shown that $f(\theta)^2$ is mathematically equivalent to the high-angle Rutherford differential cross section.

3.8. THE ORIGIN OF $f(\theta)$

Since $f(\theta)$ relates to the amplitude of a scattered wave, we will briefly consider how it arises. The following analysis is not intended to be completely rigorous, but only to give the fundamental ideas behind the meaning of $f(\theta)$ and its relation to the differential scattering cross section. You can safely delay studying this topic until curiosity wins.

To find the total elastic scattering cross section, we have to integrate $d\sigma/d\Omega$. We note that this is a particle model, but we should note how the wave nature of the electrons is brought in. We can consider the wave nature by looking at Figure 3.6.

The incident beam can described as a wave of amplitude ψ_0 and phase $2\pi kr$

$$\psi = \psi_0 \, e^{2\pi i\, kr} \qquad [3.11]$$

In this definition of phase, *k* is the magnitude of the wave vector and *r* is the distance that the wave has propagated, as we'll discuss in detail in Chapter 11. When it is scattered

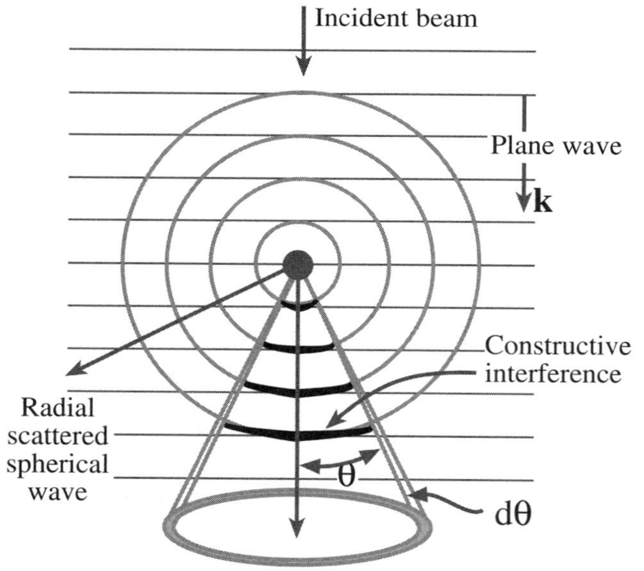

Figure 3.6. The generation of a scattered wave by the interaction of a plane wave (horizontal lines) with a point charge. The circles represent the scattered spherical wavefronts which are in phase (same λ). The in-phase constructive interference between the plane and spherical waves is shown by the dark arcs. The angles θ and $d\theta$ are the same as in Figure 2.3.

by the point charge, a spherical scattered wave is created which has amplitude ψ_{sc} but the same phase

$$\psi_{sc} = \psi_0 \, f(\theta) \, \frac{e^{2\pi ikr}}{r} \qquad [3.12]$$

In this equation, $f(\theta)$ is the amplitude we would have if $\psi_0 = 1$, i.e., it is the *atomic scattering amplitude.*

Now we need to know the atomic scattering amplitude. Up to this point, our treatment has been quite rigorous. So we need a model for $f(\theta)$ to make the problem manageable. Ideally, the model would distinguish between a neutral atom in a metal, a covalently bonded atom, and an ion. In practice, we usually use a simple approximation.

The quantity $f(\theta)$ can always, in principle, be calculated from the Schrödinger equation. If we write down the expression for the scattering process shown in Figure 3.6, then we have

$$\psi_{sc} = \psi_0 \left[e^{2\pi i \mathbf{k}_I \cdot \mathbf{r}} + i\, f(\theta) \frac{e^{2\pi ikr}}{r} \right] \qquad [3.13]$$

Note that, as usual for Huygens wavelets, there is a 90° phase shift (shown by the inclusion of i in the second term) between the incident and scattered beams, and second, that $f(\theta)$ can be expressed as

$$f(\theta) = |f(\theta)|\, e^{i\eta(\theta)} = |f(\theta)| \left(\cos \eta(\theta) + i \sin \eta(\theta) \right) \qquad [3.14]$$

which means that the phase, $\eta(\theta)$, of $f(\theta)$ also depends on θ.

> *First aside:* In writing equation 3.13, we have introduced two wave propagation parameters: the vector $\mathbf{k}_I$ for the incident plane wave and the scalar k for the spherical scattered wavelet. By writing the 2π factor separately as part of the phase term, we have implicitly defined k to be $1/\lambda$. Many physics textbooks include the 2π in k, so they have k given by $2\pi/\lambda$. Just be careful when you compare similar formulas in two textbooks.

3.9. THE STRUCTURE FACTOR *F*(θ)

The next introductory step in discussing electron scattering is to take the idea of individual atoms scattering electrons (the atomic scattering factor), which we've just discussed in some detail, and consider what happens when the atoms are stacked together in crystals. We will deal with this point in great detail in Chapter 13, but for now we can introduce the structure factor $F(\theta)$, which is a measure of the amplitude scattered by a unit cell of a crystal structure. Because it is an amplitude like $f(\theta)$ it also has dimensions of length. We can define $F(\theta)$ in terms of the sum of the atomic scattering factors from all the i atoms in the unit cell (with atomic coordinates x_i, y_i, z_i) multiplied by the phase factor that takes account of the difference in phase between waves scattered from atoms on different planes with Miller indices ($hk\ell$). The scattering angle θ is the angle between the incident and scattered electron beams. So we can write

$$F(\theta) = \sum_i f_i \, e^{i\varphi_i} = \sum_i f_i \, e^{2\pi i \left(hx_i + ky_i + \ell z_i\right)} \qquad [3.18]$$

All this means is that the amplitude (and hence the intensity) of scatter is influenced by the type of atom ($f(\theta)$), the position of the atom in the cell (x,y,z), and the specific atomic planes ($hk\ell$) that make up the crystal structure. None of this is very surprising, but it turns out that this equation predicts that in certain circumstances the amplitude of scatter is zero, which is often a very useful diagnostic test when determining crystal structures in the TEM. We'll return to this in Chapter 13 in much more detail.

> *Second aside*: The 90° phase change for the scattered wave component in equation 3.14 can be easily understood by considering the following. If the amplitude of the wave is initially $\psi_0 \sin(2\pi kr)$, then after it has passed through the specimen it will be ψ_{tot}; after scattering, the phase is increased by ϕ. We can express the new ψ_{tot} as
>
> $$\psi_{tot} = \psi_0 \sin(2\pi kz + \phi) = \psi_0 \sin(2\pi kz) \cos \phi + \psi_0 \cos(2\pi kz) \sin \phi \qquad [3.15]$$
>
> If ϕ is small, then $\cos \phi \approx 1$ and $\sin \phi \approx \phi$; $\cos \theta$ is always the same as $\sin(\theta + \pi/2)$, hence
>
> $$\psi_{tot} = \psi_0 \sin(2\pi kz) + \psi_0 \, \phi \sin\left(2\pi kz + \frac{\pi}{2}\right) \qquad [3.16]$$
>
> As Reimer notes, the $\pi/2$ would arise if we used the exponential rather than the sine to denote the phase, so we can then write equation 3.16 as
>
> $$\psi_{tot} = \psi + i\,\psi_{sc} \qquad [3.17]$$
>
> which we see then has the form given in equation 3.13.

3.10. SIMPLE DIFFRACTION CONCEPTS

As we mentioned earlier, electron diffraction is by far the most important scattering phenomenon in the TEM. The importance arises because, as we'll show you in Chapters 11 and 12, we can use diffraction to determine the spacing of planes in crystals. The interplanar spacings in different crystal structures are characteristic of that structure. As a result we can distinguish between different crystal structures by observing and measuring diffraction patterns. We'll see that the *positions* of the diffracted beams of electrons are determined by the size and shape of the unit cell, and the *intensities* of the diffracted beams are governed by the distribution, number, and types of atoms in the specimen. We'll also show you how the diffraction process leads to contrast in TEM images which is related to the orientation of a crystalline specimen.

> The combination of the diffraction pattern and the image is a most powerful tool for characterizing crystals and their defects.

It's easy to see, in a qualitative manner, how diffraction modifies the distribution of the low-angle scattering, described by $f(\theta)$, and shown for a single atom in Figure 3.5. When we consider the effect of the arrangement of atoms in the specimen, then Figure 3.5 has to be modified. For an amorphous specimen, the atoms are almost (but not quite) randomly arranged. A random arrangement would result in a similar plot as for Figure 3.5, but there are certain interatomic spacings that tend to occur in an amorphous structure (e.g., first- and second-nearest neighbor spacings are usually relatively well defined). As a result, the amplitude (and hence the intensity) of diffraction is stronger at some angles than at others, which we see as rings of intensity shown in Figure 3.7A, and in the diffraction pattern in Figure 2.11a. If the specimen is crystalline, then the intensity of the diffracted beams is a maximum at specific angles because the interplanar spacings are fixed (Figure 3.7B). The variation of $f(\theta)$ with θ in Figures 3.7A and B is equivalent to the intensity variation across the diffraction patterns in Figures 2.11a and b/c, respectively, and thus emphasizes the strong relationship between $f(\theta)$ and diffracted intensity. We'll describe this important relationship mathematically in Section 3.10.B below.

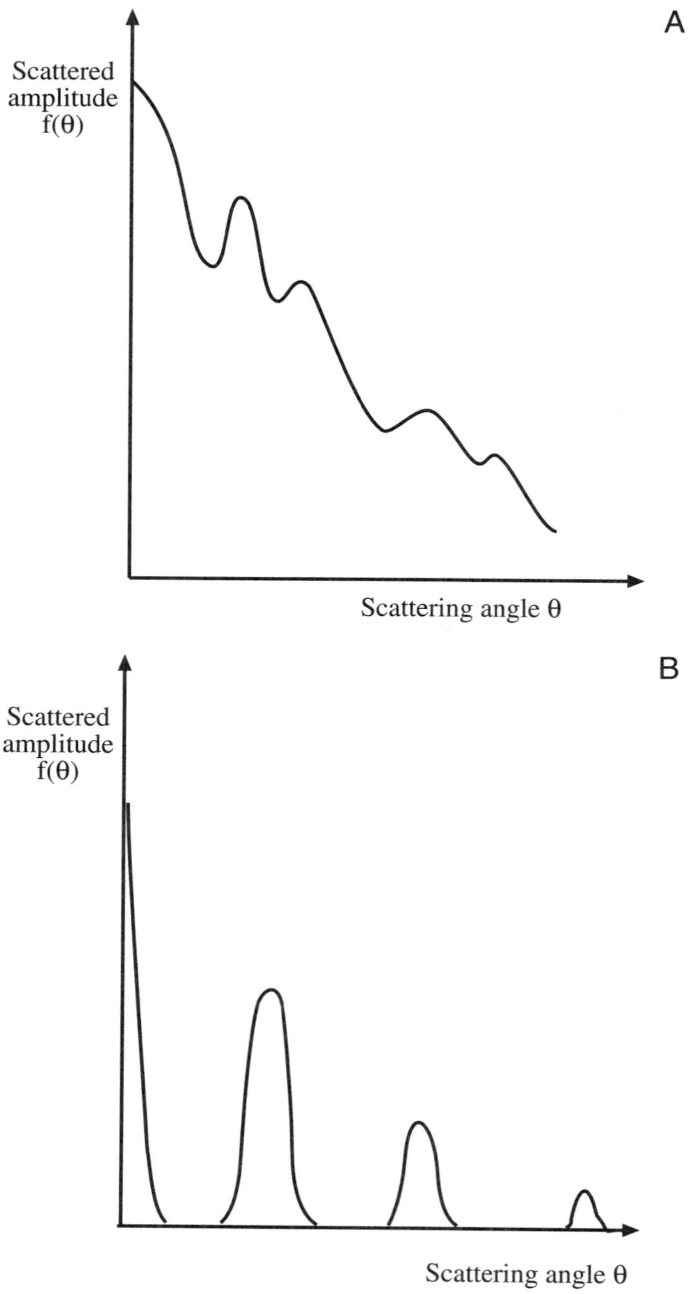

Figure 3.7. Change in $f(\theta)$ with θ (A) for an amorphous specimen and (B) for a crystalline specimen. The amplitude (and therefore the intensity) of scatter generally decreases with increasing θ, but the smooth decrease is modified at certain scattering angles (compare with the intensity variation along a line from the middle of the diffraction patterns in Figures 2.11a and b).

3.10.A. Interference of Electron Waves

To interpret this low-angle elastic scattering (which is primarily from the electron cloud) it is best to think in terms of electron waves and not in terms of particle interactions that characterize high-angle scattering. If you go back and look at Figure 3.2, you see a periodic one-dimensional array of scattering centers (slits); a monochromatic wave (fixed λ) is advancing toward the centers. Each slit acts as a new source of a wave of the same λ. Thus many new waves are created and, when more than one wave is present, the waves can interfere with one another. This process happens from even the thinnest specimens, and is entirely a wave phenomenon that doesn't need concepts such as cross section, which we apply when we think of the electron as a particle.

A rule of wave theory is that waves reinforce one another (this is constructive interference) when they are in phase, i.e., when they are coherent. Waves also cancel one another (destructive interference) when they are out of phase. What you see in Figure 3.2 is that the diffracted waves are in phase with one another only in certain directions. There is invariably a *zero-order wave* which proceeds in the same direction as the incident wave, which in the TEM we'll refer to as the direct beam of electrons. There are also *higher-order waves* which propagate in directions that are at some fixed angle to the incident wave and we'll call these the diffracted beams.

So diffraction creates many electron beams traveling at specific angles relative to a single monochromatic incident beam. In the chapters on diffraction, we'll find ways to measure these angles and relate them to the spacing of the scattering planes.

3.10.B. Diffraction Equations

Here we'll introduce the mathematical relationships that describe the diffraction process. The idea of using diffraction to probe the atomic structure of materials was accredited to von Laue in Germany in 1912, although others such as Ewald were working on similar ideas at the same time. At von Laue's suggestion, his colleagues Friedrich and Knipping irradiated a copper sulfate crystal and became the first to observe diffraction from crystal planes. In fact, it was a remarkable stroke of luck that the $CuSO_4$ diffracted the X-rays at all because of the strict equations which govern diffraction.

Von Laue used the well-known light-optics approach to argue that the diffracted waves are in phase if the path difference between waves scattered by adjacent scattering centers is a whole number of wavelengths, $h\lambda$, and h is an integer. Thus, as shown in Figure 3.8, if the scattering centers (B and C) are spaced some distance a apart and the incident beam (wavelength λ) makes an angle θ_1 with the line connecting the scattering centers, and is diffracted at an angle θ_2, then the path difference (AB – CD) is then

$$a(\cos\theta_1 - \cos\theta_2) = h\lambda \quad [3.19]$$

Now in three dimensions, two more Laue equations can be written for two more distances, b and c, and appropriate angles θ_n

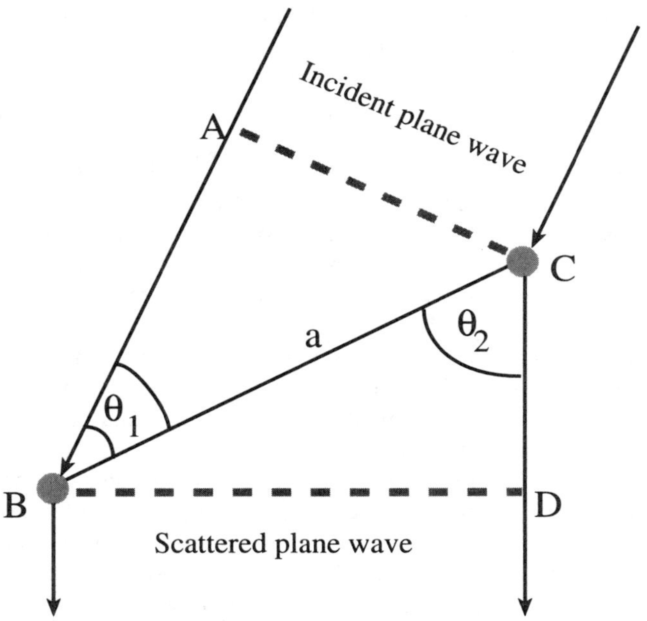

Figure 3.8. The approach used by von Laue to calculate the path difference for a wave (wavelength λ). In this one-dimensional figure the wave is incident at angle θ_1 and scattered at angle θ_2 from two atoms (B and C) spaced distance a apart. The path difference between scattered waves is AB – CD.

$$b(\cos\theta_3 - \cos\theta_4) = k\lambda \quad [3.20]$$

$$c(\cos\theta_5 - \cos\theta_6) = \ell\lambda \quad [3.21]$$

These three simultaneous equations bear von Laue's name, and for this work he received the Nobel Prize. If, in a TEM specimen, all three Laue equations are satisfied simultaneously, we will show in Chapter 11 that a diffracted beam is produced. We'll also show you in Chapters 11 and 12 that the letters $hk\ell$ are the indices of the diffracted beam and are equivalent to the Miller indices $(hk\ell)$ of a crystal plane, or some multiple thereof.

Usually in TEM, we use a simpler approach to describe diffraction. Von Laue's approach was simplified by Bragg (the elder) in England who argued that the waves behaved as if they were reflected off atomic planes as shown in Figure 3.9. Bragg showed that waves reflected off adjacent scattering centers must have a path difference equal to an integral number of wavelengths if they are to remain in phase. So, in the TEM, the path difference between electron waves reflected from the upper and lower planes in Figure 3.9 is (AB + BC). Thus, if the "reflecting" $hk\ell$ planes are spaced a distance d apart and the wave is incident and reflected at an angle θ_B, both AB and BC are equal to $d \sin \theta_B$ and the total path difference is $2d \sin \theta_B$. Then we have Bragg's law

$$n\lambda = 2d \sin \theta_B \qquad [3.22]$$

We'll reserve θ_B for the Bragg angle, which is the most important scattering angle in TEM and you'll come across it many more times in this text. Bragg also received a Nobel Prize for this one equation, even though the idea of reflected electrons, while mathematically correct, is physically wrong! We'll continue to use the term "Bragg reflection" to describe diffraction because everyone does so, even though it's inaccurate. We will demonstrate, in a rigorous fashion, the equivalence of the Bragg and Laue approaches in Chapter 12.

It is simple to see from the Bragg equation that atomic planes which are closer together give rise to larger angles of scatter. This reciprocal relationship (d is proportional to $1/\theta$; see Chapter 12) is very important in diffraction pattern interpretation. So, if we know λ for the incident electron and we can measure θ experimentally, we can work out the interplanar spacings in the crystal. It is this crystallographic information that makes diffraction such an important aspect of the TEM.

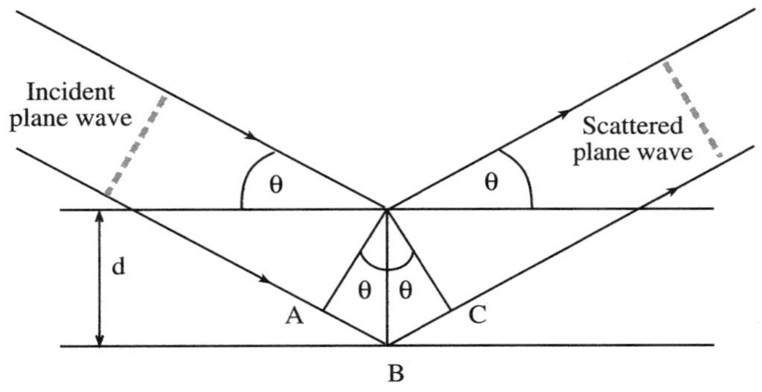

Figure 3.9. The Bragg description of diffraction in terms of the reflection of a plane wave (wavelength λ) incident at an angle θ to atomic planes of spacing d. The path difference between reflected waves is AB + BC.

CHAPTER SUMMARY

What should you remember from this chapter? Until you have time to study this material very carefully you may find it difficult, so here are a few suggestions:

- Know the words! In particular, we can describe the scattering process by three parameters:

$\sigma(\theta)$	the scattering cross section
$\dfrac{d\sigma(\theta)}{d\Omega}$	the differential scattering cross section
$f(\theta)$	the atomic scattering amplitude

In particular, don't be put off because "differential scattering cross section" sounds difficult. All three terms are *very* important in different parts of TEM.

- The relationships between $f(\theta)$ and $\sigma(\theta)$ are very important.
- The relationships between $f(\theta)$ and the intensity in a diffraction pattern are very important.

Remember that although we often write $\sigma(\theta)$ as σ, there is an angle involved in any σ.

- The fact that the electron is a charged particle is critical to the whole scattering process.

Yes, a really rigorous treatment of scattering would take into account the wave nature of the electron (wave mechanics), relativity, and the electron charge at the same time. We won't do this! Fortunately we can do very well using compiled tables.

- The strength of the scattering, $f(\theta)$, depends inversely on the scattering angle, θ.
- $F(\theta)$ is a measure of the amplitude scattered by a unit cell.

A final point to think about: remember that $f(\theta)$ is the property of a "scattering center." We usually think of this center as being an atom. What happens if the scattering center is an ion (i.e., if it is charged)? Is the scattering process affected by how this atom is bonded to its neighbors? What changes if the atom is covalently bonded? We'll answer these questions as we go on.

REFERENCES

General References

Andrews, K.W., Dyson, D.J., and Keown, S.R. (1967) *Electron Diffraction Patterns,* The Institute of Physics, Bristol, United Kingdom.

Henoc, J. and Maurice, F., Eds. (1976) *Use of Monte Carlo Calculations,* NBS Special Publication 460, Washington, DC.

Kyser, D., Ed. (1984) *Electron-Beam Specimen Interactions for Microscopy, Microanalysis and Lithography,* SEM Inc., AMF O'Hare, Illinois.

Specific References

Doyle, P.A. and Turner, P.S. (1968) *Acta Cryst.* **A24**, 390.

Hall, C.E. (1953) *Introduction to Electron Microscopy,* p. 229, McGraw-Hill, New York.

Joy, D.C., Romig, A.D. Jr., and Goldstein, J.I., Eds. (1986) *Principles of Analytical Electron Microscopy,* p. 6, Plenum Press, New York.

Reimer, L. (1993) *Transmission Electron Microscopy,* 3rd edition, Springer-Verlag, New York.

Rutherford, E. (1911) *Phil. Mag.* **21**, 669.

Taylor, C. (1987) *Diffraction*, Adam Hilger, Bristol, United Kingdom.

Inelastic Scattering and Beam Damage

4

CHAPTER PREVIEW

In the previous chapter, we discussed elastic scattering of the electron beam in which the incident electron lost no energy as it interacted with the specimen. Inelastic or energy-loss electrons are equally important and we'll discuss the processes here, but leave the applications till later. Why are we interested in inelastic scatter? Well, inelastic scattering generates a whole range of signals, each of which can tell us more about the specimen than we can find out from the elastic electrons. The most important signals are the X-rays, inelastic electrons, and secondary electrons, and so we'll emphasize how these signals arise. We will also discuss why these specific signals are useful to materials scientists.

So how do we use these other signals? First we have to detect the electrons and X-rays and we'll describe electron detection in Chapter 7. In Chapter 31 we will explain how we use some of the signals to form images of the specimen. We will discuss how to detect X-rays and get information from the spectra that are created in Chapters 32–36. Then in Chapters 37–40 we'll talk about detecting and analyzing the electrons that lose energy when they are scattered in the specimen. In all cases we get complementary information to that gained in conventional TEM images and diffraction patterns. Obviously then there's a lot of useful information in these signals and this is a major advantage to using ionizing radiation. However, the other side of the coin is that all the inelastic processes deposit energy in the specimen which can damage beam-sensitive specimens. So we must also look at the down side of the signal-generating process and we end the chapter by discussing this problem under the general topic of beam damage or radiation damage.

A warning: This chapter contains some quite difficult theoretical concepts. However, it does form the basis of AEM, which is the topic of much of Part IV of the book. You can safely delay studying much of this material in detail until you reach Chapter 32.

Inelastic Scattering and Beam Damage

4

4.1. WHICH INELASTIC PROCESSES OCCUR IN THE TEM?

Historically, the conventional TEM only used two *elastic* signals, namely, the direct beam and the diffracted beams. As we've seen, these signals constitute the diffraction pattern and we'll see in due course how they can be used to produce images. In operating a TEM in this classical manner we are being extraordinarily inefficient; we throw away a vast amount of information about our specimen which is contained in the signals that result from *inelastic* scatter. Some of those were shown back in Figure 1.3. These signals are sometimes sought in related instruments such as the SEM and the Auger electron spectrometer (AES), but we can also use TEMs to detect many of these signals, thus allowing for a more complete characterization of the specimen.

Because some of the beam electrons lose energy, all these signals are related to the general topic of electron energy-loss spectrometry (EELS). The EELS signals and the accompanying X-ray signal constitute analytical electron microscopy (AEM), which we discuss in detail in Part IV of this book. In seeking to detect more signals from the specimen, we find that practically we cannot do everything at once, nor can we do it all with equal efficiency. Nevertheless various analytical TEMs exist which, in one form or another, can detect all the signals shown in Figure 1.3. So in this chapter we'll cover all the signals that are detectable and what use (if any) they are to the materials scientist. We need to know:

- What are the inelastic scattering interactions?
- What is the energy loss associated with each process?
- What is the likelihood that each energy-loss process will occur?

When the high-energy electron encounters an atom, it first penetrates the outer, loosely bound electron cloud, then it passes the inner, more tightly bound core shell electrons, and finally it may encounter the nucleus.

> *A general rule of thumb:* The deeper the electron penetrates into the atom, the higher the energy that may be lost. It is possible (but very rare) for the electron to lose all its energy in a single nuclear interaction.

This range of inelastic scattering produces a range of scattering angles, but there is no simple relationship between the energy lost and the scattering angle. We'll separate the inelastic processes into three components:

- Processes that generate X-rays.
- Processes that generate other (secondary) electrons.
- Processes that result from collective interactions with many atoms.

We know the first two rather well, but the third is usually poorly defined. Figure 4.1 shows the cross sections for some of the more important inelastic processes that we'll talk about. These cross sections vary over several orders of magnitude and this fact alone should give you some feel for the relative generation probability of each signal. We'll discuss the specific cross sections for inelastic scatter in more detail as we describe each individual inelastic event.

4.2. X-RAY EMISSION

We'll consider X-ray emission first because it's the most important secondary signal generated in the specimen.

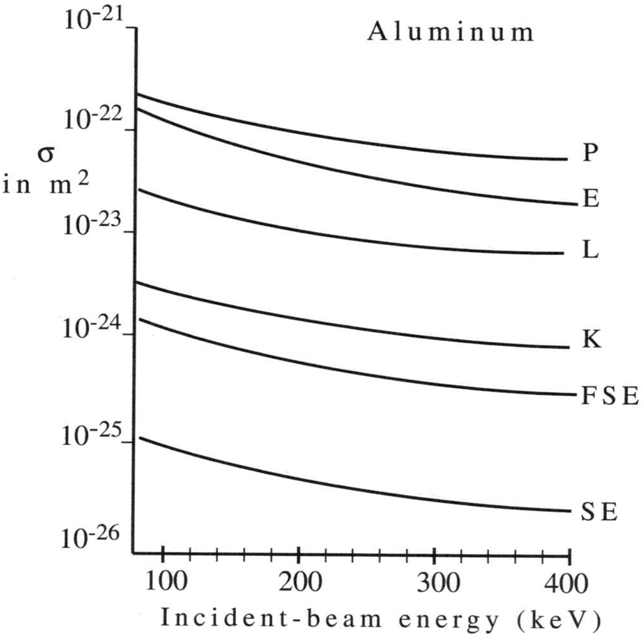

Figure 4.1. Cross sections for the various inelastic scattering processes in Al as a function of the incident electron energy, assuming a small angle of scatter ($\theta \sim 0°$); plasmon (P), K and L-shell ionization (K, L), fast and slow secondary electron generation (FSE, SE). For comparison purposes the elastic cross section (*E*) is also included. The values are relatively insensitive to the beam energy.

From X-rays we can find out easily what elements constitute the part of the specimen interacting with the electron beam and we can also quantify the amount of each element in quite a straightforward manner. (The way to do all of this is described in Part IV.) Two kinds of X-rays are produced:

- Characteristic X-rays which are useful to the materials scientist.
- Bremsstrahlung X-rays which are useful to the biologist but generally regarded as a nuisance by most materials scientists.

4.2.A. Characteristic X-rays

How do we produce characteristic X-rays and of what are they "characteristic"? First of all, a high-energy beam electron must penetrate through the outer electron shells and interact with the inner-shell (or core) electrons. If more than a critical amount of energy is transferred to an inner-shell electron, that electron is ejected; that is, it escapes the attractive field of the nucleus, leaving a hole in the inner shell. When this happens the atom is left in an excited state because it has a higher energy than it would like, and we describe it as "ionized."

The ionized atom can return almost to its lowest energy (ground state) by filling in the missing electron with one from the outer shells. It is this transition which is accompanied either by the emission of an X-ray or an Auger electron. This latter process was first described by the Frenchman Pierre Auger in 1925 and won him the Nobel Prize for Physics. Since the discoverer was French, we pronounce his name to sound like "Ogay" with a soft *g* as in "beige." In either case the energy of the emission is *characteristic* of the difference in energy of the two electron shells involved and so is unique to the atom. The process of X-ray emission is shown schematically in Figure 4.2. We'll cover Auger emission in Section 4.7.

Note that characteristic X-rays can also be produced if an atom is ionized by a process other than electron irradiation. For example, ionization can occur as a result of X-ray bombardment also, in which case we use the term "fluorescence." It is customary *not* to refer to electron-induced X-ray emission as fluorescence, although you occasionally come across such usage in the literature.

We've been able to detect X-rays in electron microscopes for many years, but Auger electron detection is rather specialized and usually carried out in a dedicated auger electron spectrometer (AES). More recently, however, we've found ways to detect the Auger signal in ultrahigh vacuum (UHV) TEMs and so we'll discuss it in Section 4.3.C below.

You need to know several aspects of the ionization process to understand why the characteristic X-ray signal is so useful and what it takes to generate it:

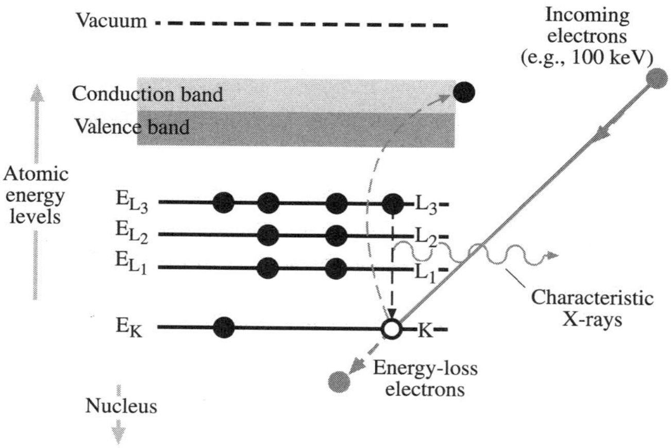

Figure 4.2. The ionization process. An inner (K) shell electron is ejected from the atom by a high-energy electron. When the hole in the K shell is filled by an electron from the L shell, characteristic K_α X-ray emission occurs. The beam electron loses energy but continues on through the specimen.

- What are *electron shells?*
- What is the *critical ionization energy* and the *ionization cross section?*
- What controls the *X-ray energy* and *wavelength?*
- What is the *fluorescence yield?*

Electron shells: We use a specific terminology to identify the different characteristic X-rays. To understand the terminology you must be familiar with the simple Bohr theory of atomic structure in which the electrons are circling the nucleus in specific shells. (The electrons stay in their shells rather than spiral into the nucleus because of the constraints imposed by quantum theory.)

> *Aside:* For historical reasons, the innermost electron shell is called the K shell and the next innermost is the L shell, and so on; we used this terminology in Figure 4.2. All the shells (except the K shell) may themselves have subshells. We name the characteristic X-rays in terms of the shell being filled and the shell from which the electron comes. (The K, L, etc. terminology was first introduced by Charles Barkla, an early X-ray spectroscopist. The reason Barkla chose K as the first shell may have been because he wasn't sure if he'd need a J shell but knew he'd need an L shell!)

Remember that the difference in the two shell energies equals the energy of the characteristic X-ray. Thus if we fill a K-shell hole from the L shell we get a K_α X-ray, but if we fill it from the M shell we get a K_β X-ray. If the hole is in the L shell and we fill it from the M shell we get an L_α X-ray, and if we fill it from the N shell we get an L_β X-ray. The notation is in fact more complex because we differentiate the α X-rays in terms of α_1 and α_2, depending from which subshell of the outer shell the electron falls to fill the hole. The α_1 X-ray is from the outermost subshell (e.g., the L_{III} or M_V), the α_2 from the next innermost (the L_{II} or M_{IV}). To make this a bit clearer you can look at the diagram in Figure 4.3. But for X-ray detection in the TEM you don't need to worry about too many details because, as you'll see later, we can't usually discriminate between the X-rays from different subshells, except at the highest X-ray energies, so K, L, and M and α and β are about all you need to remember. Much more detail is given in books on X-rays and X-ray spectrometry, e.g., Williams (1990).

Not all electron transitions are equally probable and this is taken into account by the "weights" of the lines which are given in Table 4.1. These weights are only important within a given K, L, or M family and not between families, because experimental conditions affect each family differently. In microanalysis we only use the most intense lines, usually the α lines (or, if we can't resolve them, we use the α and β lines). This will become more obvious when you've learned about X-ray qualitative analysis in Chapter 34.

Critical ionization energy: The electron beam has to transfer an amount of energy greater than a critical value to the inner-shell electron to ionize the atom. This energy is called the critical ionization energy (E_c); if we're going to generate X-rays, then the beam energy E_0 must be greater than E_c. The value of E_c increases as the electrons are more tightly bound to the nucleus, so the innermost shell (K) has a higher E_c than the L shell, and so on. Atoms with higher Z have more protons and therefore have a higher E_c. You can see this effect if you go and look at Figure 1.4, in which the energy of the X-ray peaks increases with increasing atomic number. Since there's a lot of shells and a lot of atoms, the list of critical ionization energies is long. For a complete list you have to find a reference text such as Bearden's Tables (Bearden 1964). Such a list is also invaluable in EELS since the E_c values correspond to the positions of peaks in the energy-loss spectrum which, as we'll see in Chapter 38, can be used to identify uniquely the presence of a particular ionized atom in the specimen.

The cross section for ionization (σ) is shown in Figure 4.1 for K and L shell electrons. It is not a strong function of energy and has a relatively large value, and so we expect to see X-rays generated in all TEMs. What we have to take into account, however, is a parameter called the overvoltage, U, which is the ratio of the beam energy E_0 to E_c. The cross section varies with U as shown in Figure 4.4, and what this figure tells you is that if E_0 is close to E_c then there isn't much chance of ionization. Usually in the TEM, E_0 is ≥ 100 keV and E_c is < 20 keV, so U is greater than 5 and the ionization cross section is pretty constant. Despite this simple behavior, there is considerable uncertainty about the absolute value of the ionization cross sections because few reliable experimental measurements have been made at TEM voltages. Most models are variations on the original expression given by Bethe (1930) which describes the total, not the differential, ionization cross section as

$$\sigma_T = \left(\frac{\pi e^4 b_s n_s}{E_0 E_c}\right) \log\left(\frac{c_s E_0}{E_c}\right) \qquad [4.1]$$

where the only new terms are n_s, which is the number of electrons in the ionized subshell, and b_s and c_s, which are constants for that shell. We are not particularly concerned with any angular variation in the ionization process. The differential form of the Bethe expression shows two features:

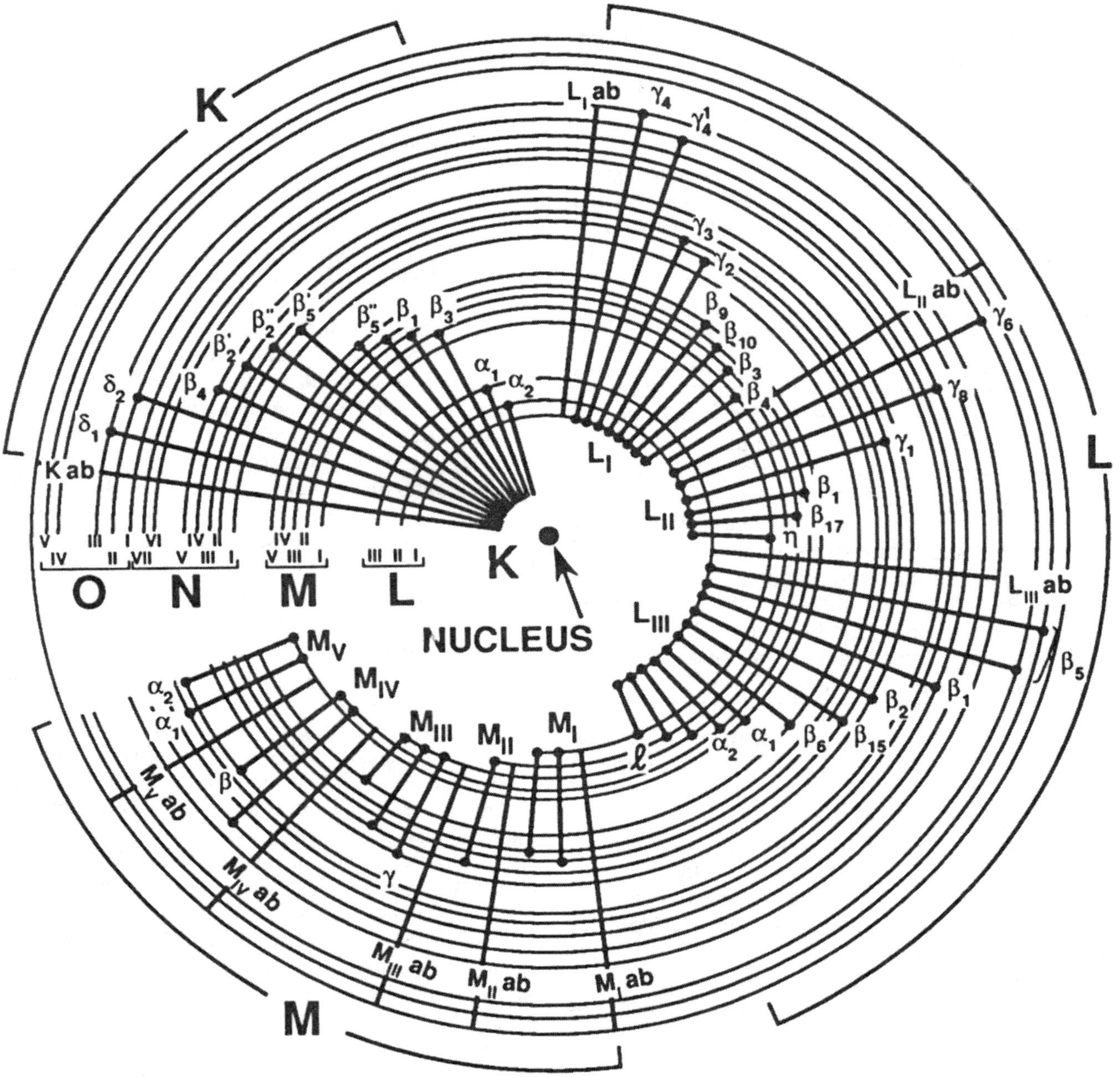

Figure 4.3. The complete range of possible electron transitions that give rise to K, L, and M characteristic X-rays. Not all these X-rays are detectable by EDS in the TEM.

- The electron that ionized the atom is deviated only through a small angle (<~10 mrads).
- The resultant characteristic X-ray is emitted uniformly over 4π sr.

As with the Rutherford cross section, the simple Bethe expression needs to be corrected for the effect of relativity at TEM electron energies, and this means substituting the term $m_0v^2/2$ for the beam energy and introducing a standard relativistic factor, β (=v/c)

$$\sigma = \left(\frac{\pi e^4 b_s n_s}{\frac{m_0 v^2}{2} E_c}\right) \log\left[\left(\frac{c_s m_0 v^2}{2E_c}\right) - \log\left(1-\beta^2\right) - \beta^2\right] \quad [4.2]$$

Table 4.1. Weights of Lines (Approximate)[a]

K_α	(1)	K_β	(0.1)		
$L_{\alpha 1,2}$	(1)	$L_{\beta 1}$	(0.7)	$L_{\beta 2}$	(0.2)
$L_{\gamma 1}$	(0.08)	$L_{\gamma 3}$	(0.03)	L_ℓ	(0.04)
L_η	(0.01)				
M_α	(1)	M_β	(0.6)	M_ζ	(0.06)
M_γ	(0.05)	$M_{II}N_{IV}$	(0.01)		

[a]The weights are given in parentheses.

This modified Bethe cross section can be manipulated to fit almost any X-ray data just by altering b_s and c_s, although such parameterization is not always justified. Several cross-section models have been developed, all of which are modifications to Bethe's approach, and Powell (1976) gives a good review.

The X-ray energy/wavelength: X-rays are electromagnetic radiation and so we usually think of them as waves with a specific wavelength λ. But just like electrons, X-rays can show particle-like characteristics and then we describe them as photons with a specific energy such as E_K or E_L, where the subscript refers to the shell from which the core electron was ejected.

There is a similar inverse relationship between the X-ray wavelength and its energy, as we saw for electrons back in Chapter 1. However, there are a couple of important differences which you should remember.

- An X-ray is a photon of electromagnetic energy, so the concepts of rest mass and momentum embodied in the electron energy are irrelevant; it has no mass.
- X-rays, like all electromagnetic radiation, travel at the speed of light (c) in vacuum and consequently we don't have to make increasing relativistic corrections as their energy increases. So the quantized X-ray energy is just $h\nu$, where h is Planck's constant and ν is the frequency, and in order to express this energy in electron volts we equate it to E, where E is the X-ray energy.

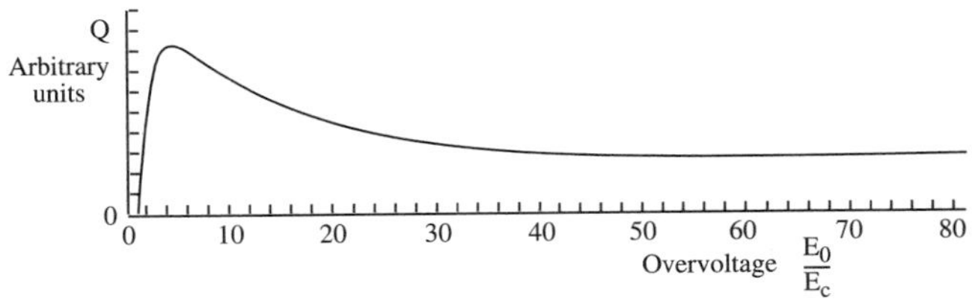

Figure 4.4. The variation of the ionization cross section with overvoltage. Ionization is most probable if the beam energy is ~5× the critical ionization energy. The cross section decreases, but not substantially, at higher overvoltages, typical of a TEM.

Thus

$$E = h\nu = \frac{hc}{\lambda} \qquad [4.3]$$

Now since h and c are constants we can substitute, and we find that

$$\lambda = \frac{1.24}{E} \qquad [4.4]$$

where λ is in nm and E in keV. This expression is *very* similar to the expression for the uncorrected electron wavelength ($1.22/E^{1/2}$, where E is in eV) that we derived back in Chapter 1 and you can easily confuse the two, so beware!

Because the X-ray energy depends on the difference in the inner-shell energies, and these differences increase monotonically with Z, we can use the detection of a characteristic X-ray with a specific energy as an unambiguous sign of the presence of an element in the specimen. The concept of the atomic number (Z) of the specimen and its relationship to the X-ray energy/wavelength was reported by the brilliant young physicist, H. Moseley (1914). Soon after his discovery, Moseley volunteered for the British army and, despite his talents, was dispatched to the trenches of Gallipoli in 1915 where he was promptly killed before he could be nominated for the Nobel Prize, which would undoubtedly have been his. He is remembered by Moseley's Law, which states

$$\lambda = \frac{B}{(Z - C)^2} \qquad [4.5]$$

where B and C are constants. So we can also generate a list of X-ray energies which are associated with each atomic transition. As with E_c the complete list is enormous and given in Bearden's Tables. More compact lists are given out in small "slide rules" by the manufacturers of X-ray spectrometers.

If you compare E_c and the X-ray energies you'll see that they are not identical. The X-ray energy E_K or E_L is invariably less than E_c. This is because the atom doesn't return completely to ground state when the X-ray is emitted. If the electron that fills the hole in the ionized inner shell comes from an outer shell, then this process will leave a hole in that shell. This hole must also be filled by another electron, and so on, until eventually a free electron from the conduction or valence band fills the last hole in one of the inner shells. So the atom returns to ground state by a cascade of transitions, depending on the complexity of the electronic structure of the atom.

An example: A Cu K shell electron requires 8.98 keV of energy for ionization (E_c = 8.98 keV). One possible sequence by which this energy is lost is first by the creation of a Cu K_α X-ray (8.048 keV), then an L_α X-ray (0.930 keV). The X-ray energies therefore total 8.978 keV and the remaining 2 eV could come from the hole in the M shell being filled from the conduction band with the emission of a photon or the generation of a phonon (see below).

The possible variations are enormous and affected by such things as Coster–Kronig transitions, in which the atomic shells rearrange their energies after the electron transition. The situation is further complicated if the ionized atom is bound to another atom, in which case the energy of the X-ray can be shifted slightly (<~5 eV). Such detail is well beyond what you need to know, but for any masochists among you the book by Dyson (1990) goes into the explicit details of this complicated subject (and our knowledge isn't complete by any means).

Fluorescence yield: Remember that an ionized atom does not have to lose energy by giving off a characteristic X-ray but can also emit an Auger electron. The probability of X-ray versus Auger emission is described by the fluorescence yield, ω, which is the ratio of X-ray emissions to inner-shell ionizations. The fluorescence yield is a strong function of atomic number as shown in Figure 4.5, decreasing at a rate proportional to Z^4 as Z decreases. One approximate expression for ω gives

$$\omega = \frac{Z^4}{a + Z^4} \qquad [4.6]$$

where $a \approx 10^6$ for the K shell. This is only an approximation but is still a formidable dependence on Z. It predicts that, for carbon ($Z = 6$), ω is ~ 10^{-3} and, for Ge ($Z = 32$), ω is ~ 0.5. This means you have to ionize 1000 carbon atoms before you get a single C K_α X-ray generated but only 2 atoms for Ge. So if you ionize low-Z atoms, the chances are you won't see an X-ray and therefore XEDS is *not* the best way to detect light elements; you should use EELS (see Part IV).

4.2.B. Bremsstrahlung X-rays

If the electrons in the beam penetrate completely through the electron shells they can interact inelastically with the nucleus. If the electron is decelerated by the Coulomb (charge) field of the nucleus, it emits an X-ray. Since the electron can suffer any amount of deceleration depending on the strength of its interaction, then these X-rays can have any energy up to the beam energy. Such X-rays produced as the electron decelerates are known by their original German name of "bremsstrahlung," which can be translated as "braking radiation."

The likelihood of bremsstrahlung creation is usually described by the cross section derived by Kramers (1923). This expression is often used for thin TEM specimens, although it was originally derived for bulk specimens. It is common to use the Kramers cross section to predict the bremsstrahlung production rather than the probability of interaction. Thus

$$N(E) = \frac{KZ(E_0 - E)}{E} \qquad [4.7]$$

where $N(E)$ is the number of bremsstrahlung photons of energy E, produced by electrons of energy E_0; K is the

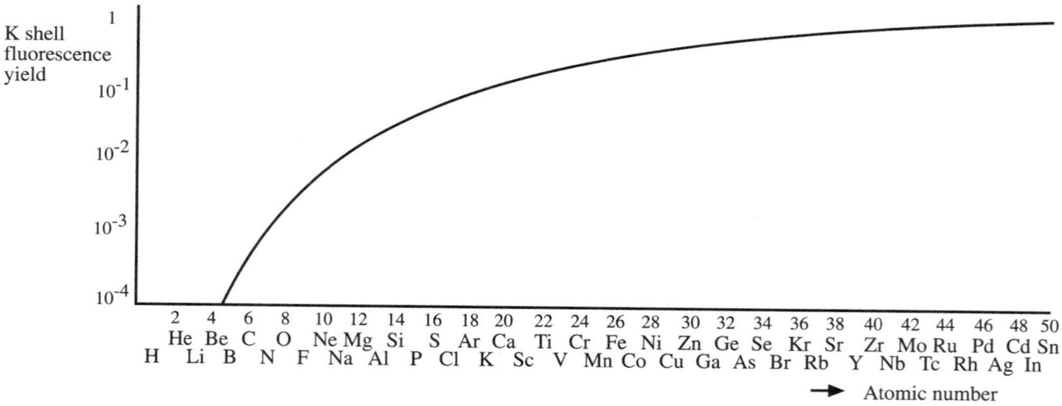

Figure 4.5. Fluorescence yield for K shell X-rays as a function of atomic number. Note the rapid decrease at low atomic numbers. X-rays from elements below Be are undetectable.

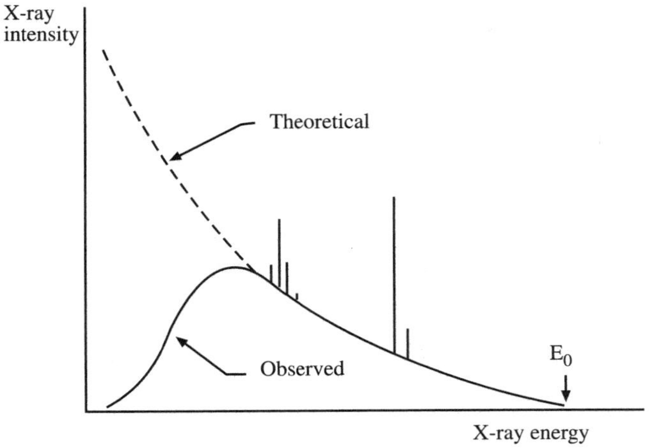

Figure 4.6. The bremsstrahlung X-ray intensity as a function of energy. The generated intensity increases rapidly at low energies but very low energy bremsstrahlung is absorbed in the specimen and the detector, so the observed intensity drops to zero. E_0 is the energy of the electrons that cause the X-ray emission. Two families of characteristic lines at specific energies are also shown superimposed on the bremsstrahlung.

Kramers constant and Z is the atomic number of the atom. This relationship predicts that it is far more likely that the braking event causes a small loss of energy and exceedingly rare that the electron loses all its energy in one deceleration at the nucleus. So the bremsstrahlung intensity is a function of energy as shown in Figure 4.6. In contrast to the isotropic emission of the characteristic X-rays, the bremsstrahlung is highly anisotropic, showing strong forward scatter which increases as E_0 increases.

The bremsstrahlung has a continuous energy spectrum on which the characteristic X-rays we just talked about are superimposed, as also shown schematically in Figure 4.6 and realistically in the experimental spectrum back in Figure 1.4. Since the characteristic X-rays have a narrow energy range, they appear as peaks in the spectrum at specific energies. The bremsstrahlung intensity depends on the average Z of the specimen and this is most useful to biologists who are interested in this aspect of their specimen. But materials scientists generally dismiss the bremsstrahlung as a useless background signal which only obscures characteristic lines. We'll come back to the X-ray spectrum in more detail in Chapter 32.

4.3. SECONDARY ELECTRON EMISSION

Secondary electrons (SEs) are electrons in the specimen that are ejected by the beam electron. They can be discussed as three distinct groups:

- If the electrons are in the conduction or valence bands then it doesn't take much energy to eject them, and they are called "slow SEs" with energies typically below about 50 eV.
- If the electrons are strongly bound inner-shell electrons they are less readily ejected, but when they are thrown out of their shells they can have a significant fraction (up to about 50%) of the beam energy, and they are then called "fast secondary electrons," or FSEs.
- If the electrons are ejected by the action of a characteristic X-ray passing its energy to an outer-shell electron, then these secondary electrons are called Auger electrons and the process is often termed a "nonradiative transition" and the X-ray is said to suffer "internal conversion" (which is not quite a religious experience).

Until quite recently SEs were only considered in relation to SEM, where they are used to form the images which are so sensitive to surface topography. We'll now discuss each of these signals and their relative importance in the TEM.

4.3.A. Slow Secondary Electrons

Slow SEs are ejected from the conduction or valence bands of the atoms in the specimen. The actual emission process can be quite complex and no simple cross-section model covers all production mechanisms. The data in Figure 4.1 indicate that SE emission is a far less likely process than all the other inelastic processes we've discussed, but enough are generated for them to be useful in the TEM. Usually, SEs are assumed to be free electrons, i.e., they are not associated with a specific atom and so they contain no specific elemental information. But because SEs are weak they can only escape if they are near the specimen surface. So we use them in SEMs for forming images of the specimen surface. While SEs are the standard signal used in SEMs, they are finding increasing use in STEMs, where they provide very high resolution topographic images of the specimen surface. We'll discuss ways to detect SEs in Chapter 7 and we'll talk about the images themselves in Chapter 31.

> SE images in a STEM have much better resolution than SE images in low-kV SEMs.

We'll discuss several reasons for this in Chapter 31. Recent developments in high-resolution field emission gun (FEG) SEMs have produced SE image resolution better than 1 nm at 30 kV, and a STEM at 100 kV can offer similar or better resolution even without an FEG, so the slow SEs are very useful. (We discuss FEGs in Chapter 5.)

The number of slow SEs with energies >~50 eV is close to zero and rises to a maximum at about 5 eV. The SE yield (number of SEs/incident beam electron) is generally regarded as being independent of E_0; if there is any Z dependence (which is still a matter of some debate) then it is very small. The angular distribution of emitted SEs is not important since the SE detector uses a strong field to gather SEs emerging from the surface at any angle. But the *number* of SEs increases with specimen tilt because SEs escape more easily as the surface is tilted parallel to the beam. This behavior is a critical aspect of SE emission because it mimics Lambert's cosine law of visible-light reflection, accounting for the great similarity between SE images of rough specimens and the reflected light images we are accustomed to seeing with our eyes.

4.3.B. Fast Secondary Electrons

Fast secondary electrons (FSEs) are high-energy electrons which are generated in the specimen; they are high-energy because they receive a large fraction of the beam energy (Joy 1984). From the cross-section data in Figure 4.1 you can see that they should be an order of magnitude more probable than slow SEs. At the low beam energies we use in an SEM, FSEs aren't a problem, so nobody bothers about them. However, in a TEM, FSEs can have energies of ~50–200 keV, in which case they not only travel significant distances within the specimen, but they may also escape from deep within the specimen. As a result, FSEs degrade the spatial resolution of microanalysis in AEMs and they also generate significant numbers of X-rays which can cause problems in quantifying X-ray data, particularly at intermediate voltages. So FSEs aren't an image resolution problem, but rather a problem for chemical analysis.

FSEs are generally both unavoidable and undesirable. We don't use them to form images or to give us spectroscopic data, but they may degrade the quality of the latter.

This phenomenon is only just beginning to be understood, but it may well turn out to be a major limitation of intermediate voltage microanalysis.

4.3.C. Auger Electrons

Remember we said at the start of this chapter that the emission of Auger electrons is an alternative to X-ray emission as an ionized atom returns to ground state. Figure 4.7 shows how an ionized atom ejects an outer-shell (Auger) electron, and it's instructive to compare with Figure 4.2 for

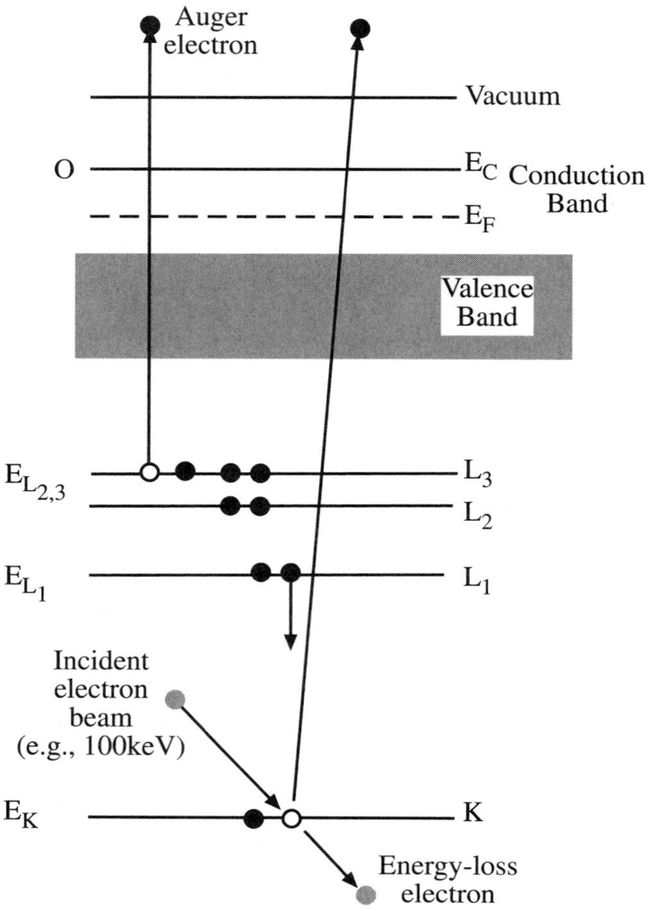

Figure 4.7. The process of inner (K) shell ionization and subsequent Auger electron emission. The energy released when the L_1 electron fills the hole in the K shell is transferred to an electron in the $L_{2,3}$ shell which is ejected as a $KL_1L_{2,3}$ Auger electron.

X-ray emission. The ejected electron has an energy given by the difference between the original excitation energy (E_c) and the binding energy of the outer shell from which the electron was ejected. So the Auger electron has a characteristic energy which is dependent on the electronic structure of the ionized atom and is almost identical to the energy of the alternative characteristic X-ray.

The Auger process is favored in atoms with a small binding energy, i.e., the lighter elements. Typical Auger electron energies are in the range of a few hundred eV to a few keV and are strongly absorbed within the specimen.

The Auger electrons that do escape come from very close to the surface. Consequently they contain surface chemical information and AES is a recognized surface

ıique. Now you might ask why light-
nalysis in the TEM is not just a surface
ıse of the similarity in energy between
and characteristic X-rays. What you have
that characteristic X-rays are much less
l in the specimen than electrons of similar
‹-rays generated in a thin TEM specimen
: detected.
.uger emission is a surface phenomenon,
ecimen surface is paramount. Oxidation
will prevent interpretable Auger analysis
chemistry and so we only carry out AES
As a result the Auger signal has tradi-
red by electron microscopists and con-
of surface chemistry, along with such
A and SIMS. However, as TEMs are be-
er vacuums and UHV STEMs become
r signal is of more interest and a few mi-
ɔorting combined Auger/STEM results.
is not simple to attach an Auger system
ıstruments are still quite rare.

ƆN–HOLE PAIRS AND IMINESCENCE (CL)

are closely related. We'll see in Chapter
letect electrons is to use a semiconductor
:tron–hole pairs when hit by high-energy
ɔur specimen happens to be a direct-gap
ɜn electron–hole pairs will be generated

› anything, the electrons and holes
, and in doing so give off light; this
ıed cathodoluminescence (CL).

ss is shown schematically in Figure 4.8.
frequency equal to the energy of the gap
Planck's constant (h), and so if the band
ne reason there will be a spectrum of light
light will vary depending on what part of
being observed. So CL spectroscopy has
ıe study of semiconductors and impurity

ɔu apply a bias to your specimen, or if it
p-n junction or a Schottky barrier diode,
s and holes can be separated under the in-
can pick up the signal if you ground the
;h a picoammeter. In this situation, the
ng as its own detector! The current you

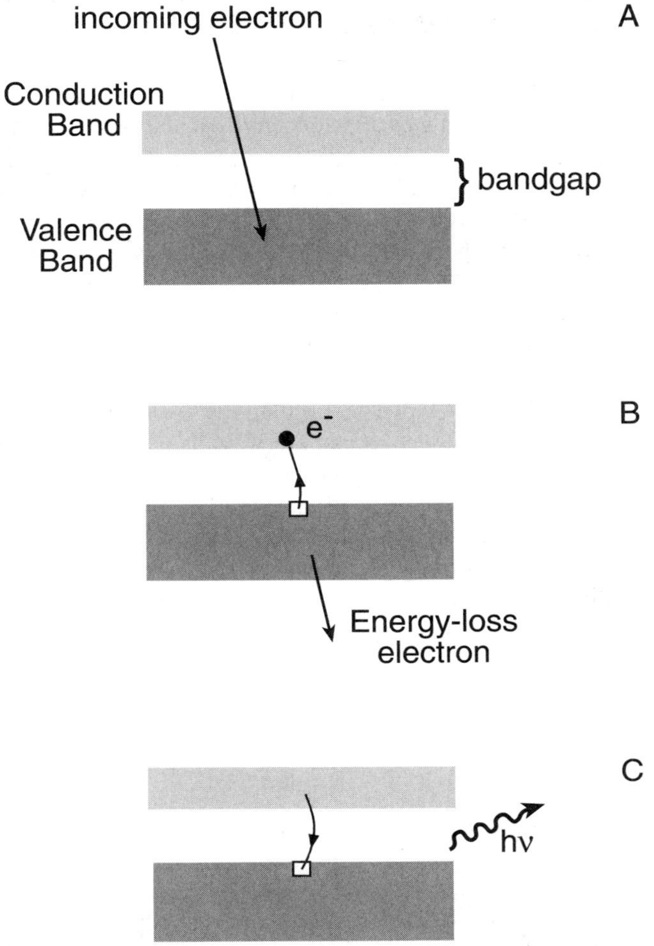

Figure 4.8. Schematic illustration of CL: (A) Initial state before a beam electron interacts with valence-band electrons. (B) A valence-band electron is excited across the gap into the conduction band, leaving a hole in the valence band. (C) The hole is filled by a conduction-band electron falling back into the valence-band hole. Upon recombination a photon of light is emitted, with a frequency determined by the band gap.

then detect is sometimes called the "electron beam induced current" or EBIC signal, and if you detect it and use it to form an image then you are doing "charge-collection microscopy" or CCM.

The CL and CCM modes of operation are standard methods of characterizing bulk samples in the SEM. In principle, there is nothing to prevent us doing the same in a STEM, and a few people have built dedicated instruments. But, in general, all these imaging modes are rare and mainly limited to studies of semiconductors (although some minerals also exhibit CL). We'll describe CL detectors in Chapter 7 and the images in Chapter 31. Just remember that CL and CCM are powerful, but rather specialized techniques.

4.5. PLASMONS AND PHONONS

We can link these two phenomena because they are both examples of what we call "collective oscillations."

> Plasmons are collective oscillations of free electrons that occur when the beam electron passes through the free electron "gas."

We can consider plasmons as analogous to sound waves, since they are longitudinal oscillations which create regions of varying electron density, as shown schematically in Figure 4.9. These oscillations are damped out in less than a femtosecond and the wave is localized to less than ten nanometers. If you go back to Figure 4.1 you'll see that the plasmon process has the largest cross section and it's by far the most common inelastic interaction occurring in materials. Plasmons can occur in any material with weakly bound or free electrons, but they occur predominantly in metals, particularly ones like aluminum which have a large Fermi surface and thus a high free-electron density. The plasmon oscillation is quantized and the mean free path for plasmon excitation is of the order of 100 nm. As we'll see in Chapter 39, this makes the number of plasmon excitations a useful way to measure the specimen thickness. Also, the plasmon energy is a function of the free-electron density and this changes with composition, so the plasmon excitation process is chemically dependent, although we rarely use it for microanalysis.

A differential cross section for plasmon excitation was given by Ferrel (1956)

$$\frac{d\sigma_\theta}{d\Omega} = \frac{1}{2\pi a_0}\left(\frac{\theta_E}{\theta^2 + \theta_E^2}\right) \quad [4.8]$$

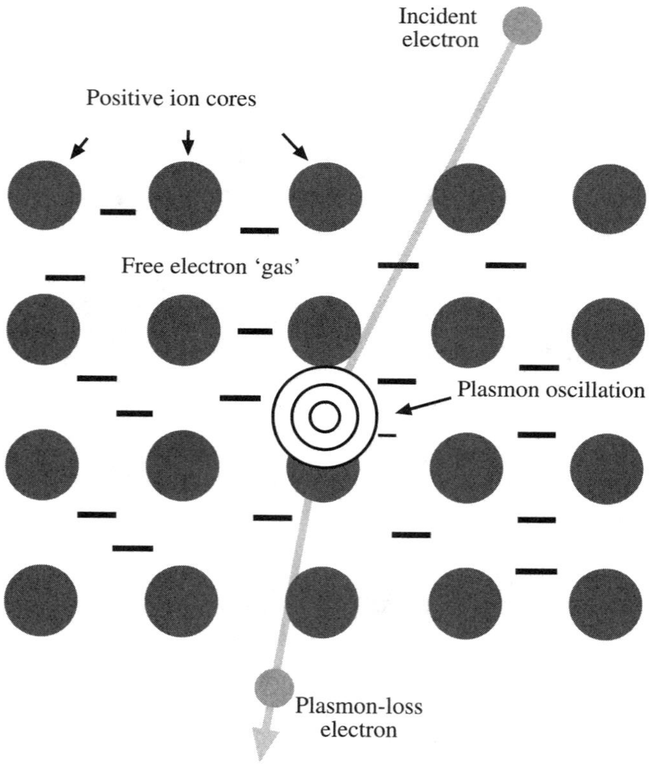

Figure 4.9. Schematic diagram of a high-energy beam electron exciting a plasmon oscillation in a free electron gas that permeates the ion cores in a metal.

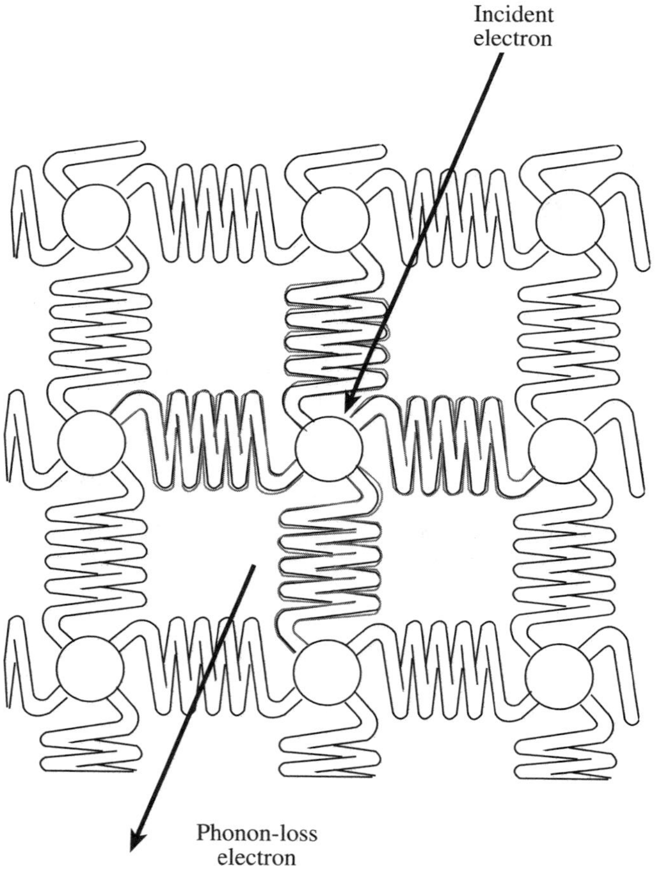

Figure 4.10. An illustration of the crystal lattice as a group of atoms linked elastically by springs. The bonds vibrate when struck by a high-energy electron creating lattice oscillations or phonons. The lattice absorbs heat by creating phonons, so phonon excitation is equivalent to heating the specimen.

where a_0 is the Bohr radius, θ is the scattering angle, and θ_E is the so-called "characteristic scattering angle" given by $E_P/2E_0$. Since E_P, the plasmon energy, is almost fixed (~15–25 eV), the cross section is a strong function of θ, dropping rapidly to zero at values much above 10 mrads, indicating once again the strong forward scattering of the electrons.

When a high-energy electron strikes an atom in the specimen, the lattice shakes, just like hitting a chain-link fence with a stick. This process occurs because, as shown in Figure 4.10, we can visualize the atoms as all linked elastically. Phonons can also be generated by other inelastic processes occurring within the atom; for example, the energy of Auger or X-ray emission or interband transitions is sometimes converted internally to lattice vibrations. Any shaking of the atoms is equivalent to heating up the specimen and the net result of all phonons is that the specimen gets warmer. As we will see, this is particularly damaging to some specimens.

The incident electron can generate phonons in any solid sample, even amorphous ones in which a crystal "structure" as such does not exist. Typically, a phonon vibration causes a very small energy loss of <0.1 eV but the phonon scattered electrons are scattered out to quite large angles (5–15 mrads), and these electrons account for the diffuse background intensity present between the Bragg spots in diffraction patterns. Phonon scattered electrons carry no useful microchemical information, nor do they carry contrast useful to the microscopist.

The phonon scattering cross section is not important to know exactly, but it is useful to remember that phonon scattering increases with Z with a dependence of approximately $Z^{3/2}$, which is rather less strong than for true elastic scattering. Also, because of the effect of temperature on atomic vibration, the phonon scatter is increased as the temperature rises. This accounts for the increase in thermal diffuse scattering with temperature, and is the major reason why we cool specimens if we want to obtain good clear diffraction patterns. The mean free path for phonon scatter at room temperature varies from a couple of nm for Au up to about 350 nm for Al, and at liquid He temperatures these values increase ~ 2–3×.

> Phonons are oscillations where all the atoms in the crystal lattice vibrate collectively. Such vibrations of the atoms are equivalent to specimen heating. You can reduce the number of phonons by cooling the specimen.

We don't use either plasmons or phonons directly to form images, but we do detect the electrons that caused them, and we'll discuss the (rather limited) uses of plasmon energy-loss electrons in Chapter 40.

4.6. BEAM DAMAGE

The inelastic collisions that give us all the useful signals we've just discussed bring with them an unfortunate side effect, that of electron beam damage. We are often less precise and call this phenomenon "radiation damage." The damage which affects the structure and/or the chemistry of the specimen depends mainly on the beam energy. Certain materials are more susceptible than others, but in the end, you can damage virtually anything that you put into the TEM. Therefore, damage represents a real physical limit on what the TEM can do and may be regarded as the microscopists' analog of the Heisenberg uncertainty principle in that the very act of observing our specimen changes it. Once the structure or the chemistry is changed, the specimen is not representative of the parent material and interpreting TEM images, diffraction patterns, and spectra becomes more difficult and eventually impossible. On the other hand, we can sometimes use beam damage to aid certain *in situ* transformations that are speeded up by the damage process or use electron damage to emulate other forms of radiation damage. Generally, however, beam damage is undesirable.

Damage takes one of two principal forms:

- *Radiolysis:* Inelastic scattering (mainly ionization) breaks the chemical bonds of certain materials such as polymers and alkali halides.
- *Knock-on damage:* Direct displacement of atoms from the crystal lattice creates point defects.

We will see that, paradoxically, the former is reduced at higher beam energies while the latter is increased, so there is sometimes no way around the problem.

Phonons represent heat in the specimen and heat is a major source of damage to polymers. Electron–electron interactions can give rise to chemical bonding changes through *radiolysis;* this process is common in polymers and alkali halides. Atomic displacement is termed "knock-on damage" within the specimen or "sputtering" if it occurs at the surface of the thin foil, and these processes are ubiquitous if E_0 is high enough. All these processes occur in the voltage range available in commercial TEMs and so you must be aware of the dangers. The actual processes can be very complicated and are also specimen-specific, so we could get bogged down in an enormous amount of detail. What we'll do is describe the fundamental processes in different materials, explain how you can determine if your specimen is being damaged, and how you can minimize or eliminate the problem. First, however, we need to know the terms we use to measure beam damage.

4.6.A. Electron Dose

In the TEM we define the electron dose as the charge density (Cm^{-2}) hitting the specimen. It is easy to convert this to the number of electrons/unit area (usually e/nm^2) knowing that $e = 1.6 \times 10^{-19}$ C. This term is *not* the same as for radiation effects on the human body, for which we define dose as the energy absorbed per unit volume; this dose is defined by the Gray (Gy), which is the absorption of 1 joule of ionizing radiation/kg of material, and 1 Gy = 100 rads (in pre-SI units). If we convert the incident electron dose to an absorbed dose it can easily be shown that typical electron exposures in the TEM are well above lethal doses for human tissue. While this is another warning about the dangers inherent in the TEM, it is more pertinent as a reminder to you that we put an enormous amount of energy into our specimens. This latter point is well illustrated if you calculate the total power input into the specimen, as we do in the next chapter. Fortunately, such a small fraction of the beam energy is transferred to a thin specimen that most specimens survive this otherwise hostile environment.

4.6.B. Specimen Heating

Specimen heating is difficult to measure experimentally because of the many experimental variables that can affect the result, such as the thermal conductivity, thickness, and surface condition of the specimen and the beam size, energy, and current. Hobbs (1979) has calculated the effects of beam current and thermal conductivity on the specimen temperature, as shown in Figure 4.11. From these results we can say that as a rule for metals and other good conductors, beam heating is negligible under standard TEM conditions, but for insulators it can be quite substantial. To minimize heating, follow the instructions given at the end of the next section.

In addition to these practical steps, beam heating is minimized by reducing the cross section for inelastic scatter, i.e., by using the highest available voltage. So HVEMs are better for the study of heat-sensitive materials. If the specimen is thinner than the mean free path for inelastic interaction, then less energy is transferred to the specimen and the result is less damage due to heating effects.

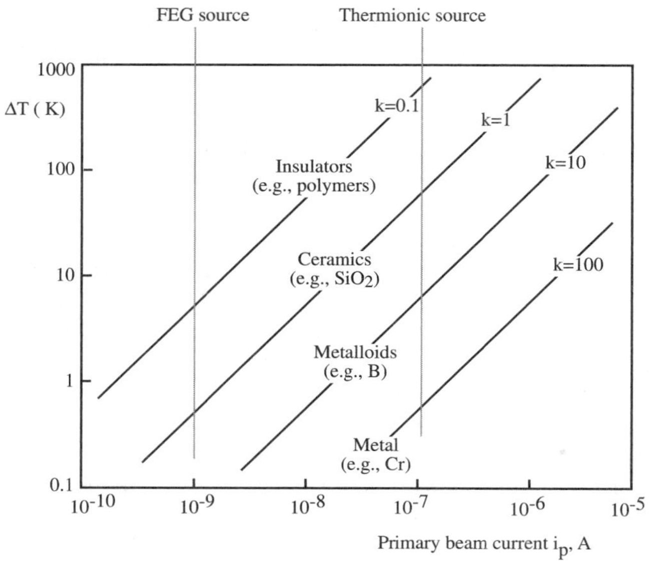

Figure 4.11. The increase in specimen temperature as a function of the beam current and the thermal conductivity k, in $Wm^{-1}K^{-1}$ of the specimen. Typical materials are noted, but should not be considered representative, since k varies substantially in any class of materials.

> Beam heating for metals is negligible under standard TEM conditions, but if thermal conduction is poor, then heating can be quite substantial. Small ceramic particles may be heated by the beam to temperatures of ~1700°C.

4.6.C. Beam Damage in Polymers

Polymers are most sensitive to the electron–electron interactions which, by one means or another, generate phonons or lattice vibrations. These phonons heat the specimen and break the bonds, creating new structures. This process is called radiolysis.

- Electrons can cause the main polymer chain to break, thus changing its basic structure.
- Electrons can cause side groups to break off, leaving reactive free radicals which may cross link to form a new structure.

A break formed this way in the polymer chain is known as scission. Generally, polymers show a tendency either to break down or to cross link. In the former case the polymer will continue to lose mass under irradiation, while in the latter the polymer eventually becomes mainly a mass of carbon. Mass loss can sometimes be measured directly in the TEM by electron spectrometry, and it can also manifest itself as major dimensional changes in the specimen.

Mass loss ultimately results in a hole forming in the illuminated area of the specimen; the image contrast will usually change before the hole appears!

If the polymer was crystalline originally, then radiation damage results in a loss of crystallinity, and you can measure this quantitatively either from the loss of diffraction contrast in the image or the loss of intensity in the diffraction pattern and the gradual appearance of an amorphous pattern. Sometimes the crystal structure can be preserved by staining. However, whenever you stain a specimen you affect its structure and mask the chemistry, so this isn't ideal.

There are several methods you can use to minimize beam damage in polymers (Sawyer and Grubb 1987):

- Use low-dose imaging techniques (see Chapter 30).
- Operate at the highest kV.
- Cool the specimen to liquid-N_2 temperatures or lower.
- Coat the specimen with a conducting metal film.
- Use STEM imaging (Section 22.3).
- Do all of the above if necessary.

4.6.D. Beam Damage in Covalent and Ionic Crystals

In covalent and ionic materials such as ceramics and minerals, radiolysis can occur which changes the specimen chemistry and possibly the structure through a series of reactions driven by the electron beam. The major inelastic interaction is that of interband transitions similar to those responsible for CL. The interband transition of a mobile valence band electron to the conduction band leaves a hole in the original energy level. Rather than emitting a photon, the electrons and holes may partially recombine via an intermediate metastable state called an exciton which, through a rather complicated sequence, can create an anion vacancy and a cation interstitial. In a similar process crystalline quartz can be amorphized. Often the process can result in the formation of new compounds which can be identified by electron diffraction and AEM. The formation of Ag from Ag halides in the photographic plate is an example of radiolysis.

We can't stop radiolysis simply by cooling or coating the specimen, since it isn't affected by heat transfer considerations. We can use higher voltages to lower the cross section for the electron–electron interactions. The best way is to use both higher voltages and thin specimens. Nevertheless, radiolysis remains a major limitation when looking at certain ceramics and minerals, and many polymers in the TEM.

4.6.E. Beam Damage in Metals

The primary way that metals are affected is by knock-on or displacement damage. This process occurs by the direct transfer of the beam energy to atoms in the solid, knocking them out of their atomic site and creating a combination vacancy and interstitial (or Frenkel pair).

> Knock-on damage is directly related to the beam energy.

How strongly the atoms are bonded to their neighbors will also be a factor. A simple expression given by Hobbs (1979) for the displacement energy E_d allows us to determine the threshold energy (E_t) for displacement of atoms of atomic weight A

$$E_t = \frac{\left(\dfrac{100 + AE_d}{5}\right)^{1/2} - 10}{20} \qquad [4.9]$$

where E_t is in MeV and E_d is in eV; E_d is typically in the range from 5–50 eV, but varies with bonding type. If we assume that a typical value of E_d is ~25 eV, we can determine the threshold potentials for a range of elements from Figure 4.12. From this figure it is quite evident that if you have a 400-kV intermediate voltage TEM, you can displace atoms with atomic weight below about Ti. If you're using an HVEM with beam energies of 1 MeV or more you will *invariably* cause displacement damage, except perhaps in the heaviest elements, or those with particularly strong covalent bonds such as diamond. The only way to avoid displacement damage is to operate below threshold.

How can you identify displacement damage? It usually manifests itself as small vacancy clusters which appear as black-white lobe contrast or dot contrast as we showed back in Figure 1.8, or sometimes damage is discernible as dislocation loops. Displacement damage can also occur in polymers and minerals, of course. The problem here is that we just suggested going to higher voltages as one way of minimizing thermal effects and radiolysis. So depending on your specimen there may in fact be no way to avoid damage of one form or another in the TEM.

The only bright side to displacement damage is that we can study it for its own sake. It can be argued, though by no means conclusively, that electron beam damage in materials can be equivalent to neutron damage, such as that

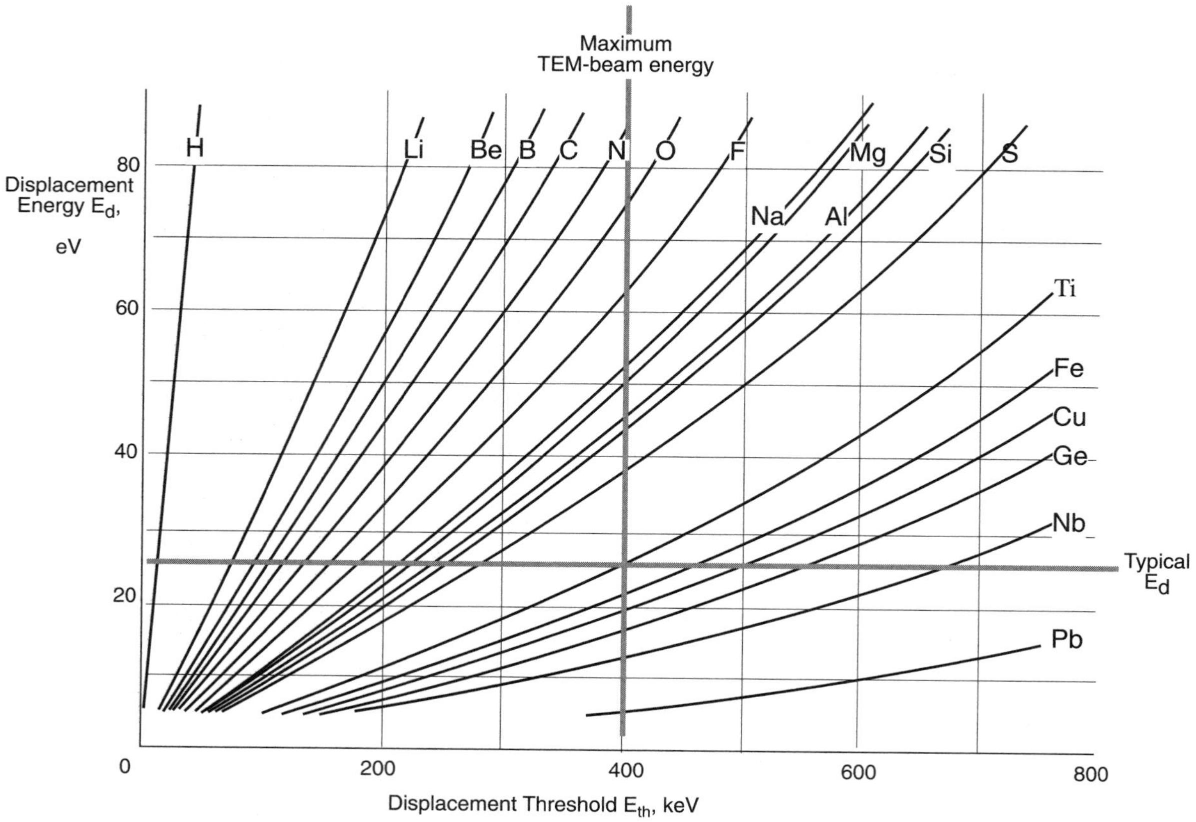

Figure 4.12. The displacement energy for a range of atoms as a function of the threshold energy (i.e., the beam energy) required for displacement damage. In a typical material E_d is ~15–25 eV, but it can vary substantially with bond strength.

Table 4.2. Comparison of Maximum Transferable Kinetic Energy with Displacement and Sputtering Energies at 100, 200, 300, and 400 kV (from Zaluzec and Mansfield 1987)

	T (eV)					
Element	100 kV	200 kV	300 kV	400 kV	*T* (eV)	*T* (eV)
Al	8.93	19.5	31.6	45.3	16	4–8
Ti	5.00	11.0	17.8	25.5	15	4–8
V	4.73	10.3	16.72	24.0	29	7–14
Cr	4.63	10.1	16.38	23.5	22	5–11
Fe	4.31	9.40	15.25	21.8	16	4–8
Co	4.08	8.91	14.45	20.7	23	5–12
Ni	4.10	8.94	14.5	20.8	22	6–11
Cu	3.79	8.26	13.4	19.2	18	4–9
Zn	3.69	8.03	13.03	18.7	16	4–8
Nb	2.59	5.65	9.17	13.2	24	6–12
Mo	2.51	5.47	8.88	12.7	27	7–14
Ag	2.23	4.87	7.90	11.3	28	7–14
Cd	2.14	4.67	7.58	10.9	20	5–10
Ta	1.33	2.90	4.71	6.75	33	8–16
Pt	1.23	2.69	4.37	6.26	33	8–16
Au	1.22	2.67	4.32	6.2	36	9–18

occurring in nuclear reactors. A general rule of thumb was that a few minutes' exposure in an HVEM was equivalent to many years in a nuclear reactor and so accelerated studies of materials degradation were possible. With this justification, an enormous amount of work was carried out in the 1960s when nuclear power was in vogue. Three Mile Island and Chernobyl have seriously reduced the number of such studies, but if you want to find out about it there are reviews in the literature, such as Laidler and Mastel (1975).

Vacancies caused by displacement damage can enhance diffusion processes, and this in turn can speed diffusional transformations when they're being studied *in situ* in the HVEM. There are many other problems that can arise when doing this, and other *in situ* observations, so interpretation isn't always straightforward. The book by Butler and Hale (1981) is recommended for more facts.

4.6.F. Sputtering

The displacement of surface atoms, or sputtering, occurs in the TEM, at voltages which are about 50% less than knock-on thresholds. If your specimen is quite thick then this problem is minor, but often the specimen has to be very thin if you want the best images and the best microanalytical resolution. In these circumstances sputtering may substantially change the surface chemistry of the specimen and affect quantitative microanalysis. Table 4.2 lists typical sputtering threshold voltages compared with displacement thresholds and, as you can see, there is cause for concern even at 100 kV.

CHAPTER SUMMARY

Inelastic scatter transfers energy to the specimen, generating a lot of useful signals which we can use to form different images of the specimen or get spectroscopic information about its chemistry and electronic structure. Unfortunately, the same processes transfer heat to the specimen which can be disastrous for certain materials such as polymers. To minimize heat transfer, higher voltages should be used, but eventually knock-on and sputtering damage occur which create defects and change the surface chemistry of all materials.

REFERENCES

General References

Inokuti, M. (1971) *Rev. Mod. Phys.* **43**, 297.

Mott, N.F. and Massey, H.S.W. (1965) *The Theory of Atomic Collisions,* Oxford University Press, New York.

Wang, Z.L. (1995) *Elastic and Inelastic Scattering in Electron Diffraction and Imaging*, Plenum Press, New York. Covers more than inelastically scattered electrons using a more rigorous mathematical approach than in this chapter.

Specific References

Auger, P. (1925) *J. Phys. Rad.* **6**, 20.

Bearden, J.A. (1964) NYO-10586, US Atomic Energy Commission, Oak Ridge, Tennessee.

Bethe, H.A. (1930) *Ann. der Phys. (Leipzig)* **5**, 325.

Butler, E.P. and Hale, K.F. (1981) "Dynamic Experiments in the Electron Microscope," in *Practical Methods in Electron Microscopy,* **9** (Ed. A.M. Glauert), Elsevier, Amsterdam.

Dyson, R. (1990) *X-rays in Atomic and Nuclear Physics,* Cambridge University Press, Cambridge, United Kingdom.

Ferrel, C.R. (1956) *Phys. Rev.* **101**, 554.

Hobbs, L.W. (1979) in *Introduction to Analytical Electron Microscopy* (Eds. J.J. Hren, J.I. Goldstein, and D.C. Joy), p. 437, Plenum Press, New York.

Joy, D.C. (1984) *J. Microsc.* **136**, 241.

Joy, D.C., Goldstein, J.I., and Romig, A.D. Jr., Eds. (1986) *Principles of Analytical Electron Microscopy,* p. 1, Plenum Press, New York.

Kramers, M.A. (1923) *Phil. Mag.* **46**, 836.

Laidler, J.J. and Mastel, B. (1975) in *Physical Aspects of Electron Microscopy and Microbeam Analysis* (Eds. B.M. Siegel and D.R. Beaman), p. 103, Wiley, New York.

Moseley, H.E.G. (1914) *Phil. Mag.* **26**, 1024.

Powell, C.J. (1976) *Rev. Mod. Phys.* **48**, 33.

Sawyer, L.C. and Grubb, D.T. (1987) *Polymer Microscopy,* Chapman and Hall, New York.

Williams, K.L. (1990) *X-ray Spectrometry,* Allen and Unwin, London.

Zaluzec, N.J. and Mansfield, J.F. (1987) in *Intermediate Voltage Microscopy and Its Application to Materials Science* (Ed. K. Rajan), p. 9, Philips Electron Optics Publishing Group, Mahwah, New Jersey.

Electron Sources

5

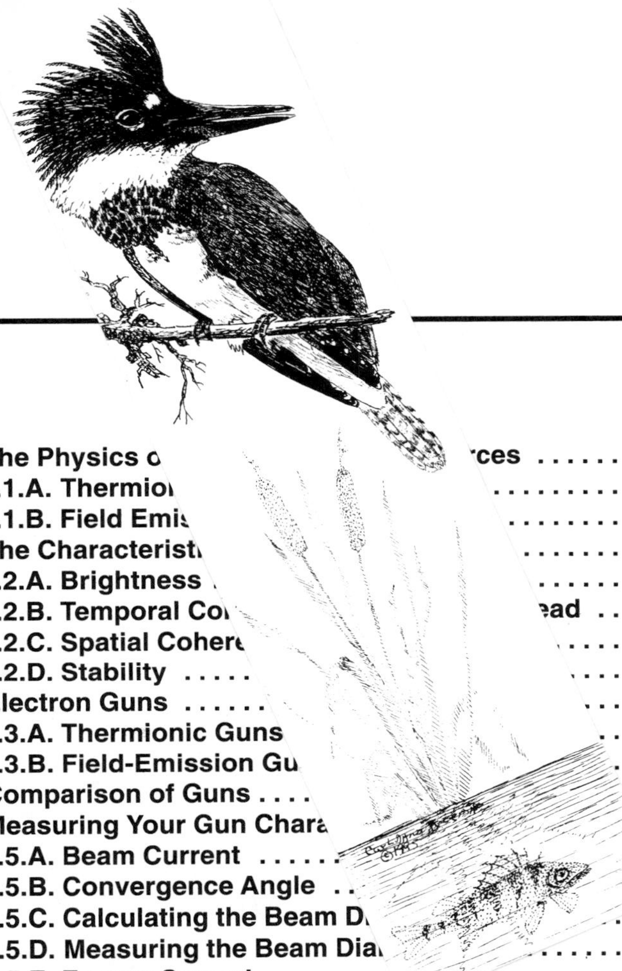

CHAPTER PREVIEW

All microscopes need a source of electrons to illuminate the specimen. Fortunately, electron sources are plentiful, but to get the best images and other signals out of our expensive microscope, we need to use the best available source. There are stringent requirements for the beam of electrons and these are best met by only two types of source: thermionic and field-emission sources. Thermionic sources are either tungsten filaments or lanthanum hexaboride (LaB_6) crystals, and field emitters are fine tungsten needles. In this chapter we'll first explain briefly the physics of these two emission processes because then you'll understand why we operate the sources in certain ways. Next we'll tell you the characteristics we need from our electron beam. Then we'll compare the three sources and show you that no one source is best for all aspects of TEM, but all three have their roles. Finally, we'll explain ways to check that a particular source meets your specification.

Because the source is so critical to the performance of the microscope, the technology is advancing rapidly to the point of complete computer control, which would leave you, the operator, with precious little to do except push the "on" button. This state of affairs is most advanced for the field-emission source, and since these are both delicate and expensive, it is just as well. But the vast majority of TEMs still use thermionic sources, and these need a fair bit of operator control. In these circumstances, you should know how these sources work and why you do certain things to them. So we'll spend most of this chapter talking about thermionic sources, although there's a good chance that field emission will be the source of choice in the future.

Electron Sources 5

5.1. THE PHYSICS OF DIFFERENT ELECTRON SOURCES

We use two kinds of electron sources in TEMs: the first kind is called a thermionic source, which, as the name suggests, produces electrons when heated, and the second type is a field-emission source, which produces electrons when an intense electric field is applied to it. These sources are part of an assembly which we refer to as the "electron gun." Now, from a physics standpoint, it is really quite interesting to know the details of how electron sources work and there's a great deal of active research into new and improved sources. However, from a practical standpoint, you don't have to know too much of the physics, and we can summarize the essential points very briefly, using a few simple equations. Keep in mind two points as you read about sources:

> Your TEM will use a thermionic source or a field-emission source and the two cannot be interchanged.
> Field-emission sources give "monochromatic" electrons; thermionic sources are less monochromatic and give "whiter" electrons.

The analogy here is to X-rays or visible light. You don't always want to use "monochromatic" electrons, even if the field-emission TEM did cost twice as much as a "conventional" microscope would with a thermionic source.

5.1.A. Thermionic Emission

If we heat any material to a high enough temperature, we can give the electrons sufficient energy to overcome the natural barrier that prevents them from leaking out. This barrier is termed the "work function" (Φ) and has a value of a few electron volts.

The physics of thermionic emission can be summarized in Richardson's Law, which relates the current density from the source, J, to the operating temperature, T in Kelvin

$$J = AT^2 e^{-\frac{\Phi}{kT}} \qquad [5.1]$$

where k is Boltzmann's constant (8.6×10^{-5} eV/K) and A is Richardson's "constant" (A/m^2 K^2), which depends on the source material. From this equation then you can see that we need to heat the source to a temperature T such that energy greater than Φ is given to the electrons; then they will escape from the source and be available to form an electron beam. Unfortunately, when we put a few eV of thermal energy into most materials they either melt or vaporize. So the only viable thermionic sources are either refractory (high melting point) materials or those with an exceptionally low work function. In practice we use both types: tungsten has the necessary high melting temperature (3660 K) and lanthanum hexaboride (LaB_6) has a low work function. If you look ahead to Table 5.1, you'll see the relative values of J_c, T, and Φ for tungsten and LaB_6.

We use several different words to describe the sources. We sometimes call tungsten sources "filaments," because tungsten can be drawn into fine "thread" which is about 0.1 mm in diameter and is similar to the filament used in an incandescent light bulb. The wire is bent into a V shape so they're also called "hairpin" filaments, or they may be sharpened to a fine point. For decades these have been the standard source in most electron-beam instruments. LaB_6, or other rare-earth boride crystals (which should not be called filaments) are usually grown with a <110> orientation to enhance emission. Sometimes we call both tungsten and LaB_6 sources "cathodes" because, as we'll see, the complete gun assembly acts as a triode system in which the source is the cathode.

Thermionic sources:	W hairpin Pointed W LaB_6 and other low-Φ materials, e.g., CeB_6

So all you need to know from the physics is that heating up a thermionic source gives you a higher J. But there is a limit because higher temperatures shorten the source life through evaporation and/or oxidation. So we seek a compromise operating temperature, and we achieve this by operating under a condition called "saturation," which we'll discuss in Section 5.3.A.

5.1.B. Field Emission

Field-emission sources operate on a fundamentally different principle than thermionic sources. The principle behind field emission is that the strength of an electric field E is considerably increased at sharp points, because if we have a voltage V applied to a (spherical) point of radius r then

$$E = \frac{V}{r} \qquad [5.2]$$

The technique of field-ion microscopy is another well established experimental tool. It requires specimens with a very fine needle shape, and so there's a lot of expertise available to help produce field-emission electron sources. One of the easiest materials to produce with a fine tip is tungsten wire, which can readily be given a tip radius of <0.1 μm. If we apply a 1-kV potential to this tip, then E is 10^{10} V/m and this lowers the work function barrier sufficiently for electrons to tunnel out of the tungsten. This process imposes quite severe stress on the tip and the material has to be strong. Field emission, like thermionic emission from LaB_6, depends on the crystallography of the tungsten tip; the <310> orientation is found to be best.

To allow field emission, the surface has to be pristine, that is, free of contaminants and oxide. We can achieve this by operating in UHV conditions ($<10^{-11}$ Torr), and in this case the tungsten is operated at ambient temperatures and the process is called "cold field emission." Alternatively, we can keep the surface in a pristine condition at a poorer vacuum by heating the tip. The thermal energy assists in electron emission so much that, in fact, the electrons don't tunnel through the barrier. For such "thermal field emission," surface treatments with ZrO_2 improve the emission characteristics, particularly the stability of the source, and such "Schottky" emitters are becoming popular. New sources such as semiconductor p-n field emitters are also provoking some interest.

5.2. THE CHARACTERISTICS OF THE ELECTRON BEAM

The electron beam in a TEM requires certain characteristics which are controlled by the source itself and how we integrate the source into a gun assembly. We describe the performance of an electron source by such terms as "brightness," "coherency," and "stability." While these words mean something to you already, they have very precise meanings in TEM terminology, so we'll go through the various characteristics, tell you what they mean, and why they are important in the TEM. We'll then compare the properties of the various sources that you may have in your microscope and you'll see that there's no "best" source for all applications, but for specific applications one source or another is clearly the best.

Before we define the electron beam characteristics needed in a TEM, it is worth summarizing here a few of the properties of electron beams in general and how these vary with accelerating voltage.

5.2.A. Brightness

The word "brightness" is often confused with "intensity" and indeed the two terms are related. For instance, when we look at the viewing screen of a TEM, we may say how "bright" it is, when we are really referring to the intensity of light coming from the screen. When we think of the intensity of any radiation source, it is in terms of the flux emanating from it. For a light bulb, it would be the number of photons per unit area per unit time. For electron sources we talk about the current density, which is the number of electrons (or charge) per unit area per unit time.

> While current density can be a useful term, it is more important to define the *brightness.* Brightness is the *current density per unit solid angle* of the source.

Electron sources differ considerably in their size and, as a result, the electrons leave the source with a range of angles, so we can't ignore the angular distribution of the electrons. Brightness is particularly important when we are using very fine electron beams, as we do in analytical and scanning microscopy. The concept of brightness is less important in conventional TEM, where we use a relatively large, defocused beam, but it is still relevant to the intensity we see on the screen, and so it affects how easy it is to operate the microscope and see our images and diffraction patterns.

So we can consider an electron source as having the following characteristics:

- a diameter d_0,
- giving off a certain cathode emission current i_e,
- the electrons diverging from the source with a semiangle α_0.

The actual way in which this is achieved we'll talk about in Section 5.3, where we discuss the complete gun assembly, but if you look ahead at Figure 5.1 you'll see that i_e, d_0, and α_0 are actually defined at the gun crossover, that is, the point at which the electrons are focused after leaving the source. The current density (current per unit area) is $i_e / \pi(d_0/2)^2$ and the solid angle of the source is $\pi\alpha_0^2$, so we define the brightness, β, as

$$\beta = \frac{i_e}{\pi\left(\frac{d_0}{2}\right)^2 \pi(\alpha_0)^2} = \frac{4\, i_e}{(\pi\, d_0\, \alpha_0)^2} \qquad [5.3]$$

> This equation is an important one which you should remember. The units of β are usually A/(cm^2 sr) or A cm^{-2} sr^{-1}. Again we see that microscopists are not comfortable using SI units, which would be A m^{-2} sr^{-1}, increasing the traditional brightness number by a factor of 10^4. What is not shown in this equation is the important fact embodied in equation 5.1 that β increases linearly with increasing accelerating voltage for thermionic sources. This is one reason for the development of intermediate voltage (300–400 kV) instruments.

Obviously, the higher the value of β, the more electrons we can put into an electron beam of a given size, and so the more information we can extract from the specimen and the more we can damage sensitive specimens. The beam current is an important part of the brightness equation. Having some way of measuring the beam current *in situ* can be a very good diagnostic tool. We'll talk about this later in the chapter when we discuss measuring the source brightness, but for the time being you can again look ahead to Table 5.1 to see how the three sources compare in brightness, which we have given in non-traditional SI units.

Now we can consider some real numbers. With a cold FEG at 100 keV, we can put 1 nA into an area of diameter 1 nm at a maximum. If you convert this current density to units of power (1 watt = 1 J/s), you'll find that the energy the electron beam puts into this small area of the specimen is nearly 150 MW/mm^2. The output of a typical electric power generating turbine is only about 600 MW. Clearly, we can change our specimen when we look at it in the TEM as we discussed in relation to beam damage. The energy density we just calculated means that an electron source is the brightest continuous radiation source known; it is considerably brighter than a supernova.

The brightness is particularly important in AEM, which is the technique of quantitative analysis of the many signals that come from a specimen irradiated by an electron beam, shown back in Figure 1.3. Similarly, as we go to higher magnifications in HRTEM, the screen intensity becomes less because we are viewing only a fraction of the illuminated area of the specimen. The electron density can be increased by using the brightest available source. Then images can be recorded with reasonably short exposure times.

5.2.B. Temporal Coherency and Energy Spread

The coherency of a beam of electrons is a way of defining how well the electron waves are "in step" with one another. You know that white light is incoherent, because it consists of photons with a range of wavelengths (colors), and so to get a coherent beam of electrons we must create one in which all the electrons have the same wavelength, just like monochromatic light. We refer to this aspect of coherency as "temporal coherency," which is a measure of how similar the "wave packets" are. If they are all identical they have the same coherence length. A definition of the coherence length λ_c is

$$\lambda_c = \frac{v\,h}{\Delta E} \qquad [5.4]$$

where v is the electron velocity, ΔE is the energy spread of the beam, and h is Planck's constant. This means we must have stable power supplies to the source and a stable high-voltage supply (or high tension, as it's sometimes called for historical reasons) so that all the electrons have a small ΔE, thus giving a well-defined wavelength. Now in practice it's impossible to create a truly monochromatic beam and we have to live with a certain range of electron energies/wavelengths, although the stability of electronic components has improved substantially over the years. Again if you look at Table 5.1 you'll see typical ΔE values for the three sources and they're in the range 0.1 to 3 eV (which is remarkably small compared with a total energy of 100 to 400 keV). So it isn't really correct to imply that thermionic sources give "white" electrons since ΔE is still small. From these values of ΔE, if you take care to get the units consistent, you can calculate typical coherence lengths, which turn out to be a few hundred nanometers.

Temporal coherency is important when the energy spread of the electrons that are *incident* on the specimen affects the microscopy. Because we can make such good high-tension power supplies, this rarely limits any aspect of TEM except perhaps high energy-resolution electron spectrometry (see Chapters 37–40). In other words, for most practical purposes our electron sources are stable enough. However, we'll see that it's a very different matter when we have to consider the electrons that have come *through* the specimen because they may have lost substantial amounts of energy.

5.2.C. Spatial Coherency and Source Size

Spatial coherency is related to the size of the source. Perfect spatial coherence would imply that the electrons were all emanating from the same point at the source. So source size governs spatial coherence and smaller source sizes give better coherency (just as they give higher brightness). The spatial coherence is strictly defined by looking at electron interference fringes in the equivalent of a Fresnel biprism experiment in light optics, with which you may be familiar. We can define the distance d_c, the effective source size, for coherent illumination to be

$$d_c << \frac{\lambda}{2\alpha} \qquad [5.5]$$

where λ is the electron wavelength and α is the angle subtended by the source at the specimen. We can control α by inserting an aperture in the illumination system, as we'll see when we describe the construction of a TEM in Chapter 9. But if this aperture is not limiting then it is the smallest source which subtends the smallest angle, and thus has the highest spatial coherence. Putting reasonable values for 100-keV electrons into equation 5.5 we find that the spatial coherence is at best only about a nanometer. To maximize the coherency, you can choose several approaches:

- Make the source size d_c smaller, e.g., by using a field-emission source.
- Use a smaller illumination aperture, thus reducing α.
- If your source size is large (e.g., a W hairpin) decrease the accelerating voltage and thus increase λ.

Spatial coherency is more important practically than temporal coherency. A small electron source subtends a small angle at the specimen, and we can help by using small limiting apertures. Small beams are more spatially coherent than large beams. The more coherent the beam is, the better the quality of the high-resolution phase-contrast images (see Part III), the sharper the diffraction patterns (see Part II), and the better the diffraction contrast in images of crystalline specimens (see Part III). An in-depth and rather mathematical description of coherency in the TEM is given in the review by Hawkes (1978).

5.2.D. Stability

In addition to the stability of the high-voltage supply to the source, it is also important that the electron current coming from the source is stable. Otherwise, the screen intensity will vary, making it difficult for you to take correctly exposed images, and also making microanalysis impossible in many cases. Thermionic sources are generally very stable except when they are first installed, or when they are about to fail. Typically, you can expect variations of less than 1% per hour in the current. For cold field emission sources, however, the emission current is not very stable, and electrical feedback circuits are required to maintain stability to better than 5%. Stability does improve with better UHV conditions.

> To summarize, the important properties of electron sources are their brightness, temporal coherency, energy spread, spatial coherency, and stability. A smaller source size gives higher β and better spatial coherency, but less stability.

Now that we know the critical characteristics required of electron sources, let's examine those used in commercial TEMs.

5.3. ELECTRON GUNS

It's no good just having a source. We need to be able to control the electron beam and direct it into the illumination system of the TEM. We do this by incorporating the source into a gun assembly which in effect acts as a lens to focus the electrons coming off the source. The design of the gun is different for thermionic sources and field-emission sources.

5.3.A. Thermionic Guns

Both tungsten and LaB_6 sources are used as the cathode in a triode gun shown in Figure 5.1. In addition to the cathode, there is a "grid" called a Wehnelt cylinder, and an anode at earth potential with a hole in its center. What these three components look like in practice is shown in Figure 5.2, where they are all separated. The cathode is attached to the high-tension cable, which in turn connects to the high-

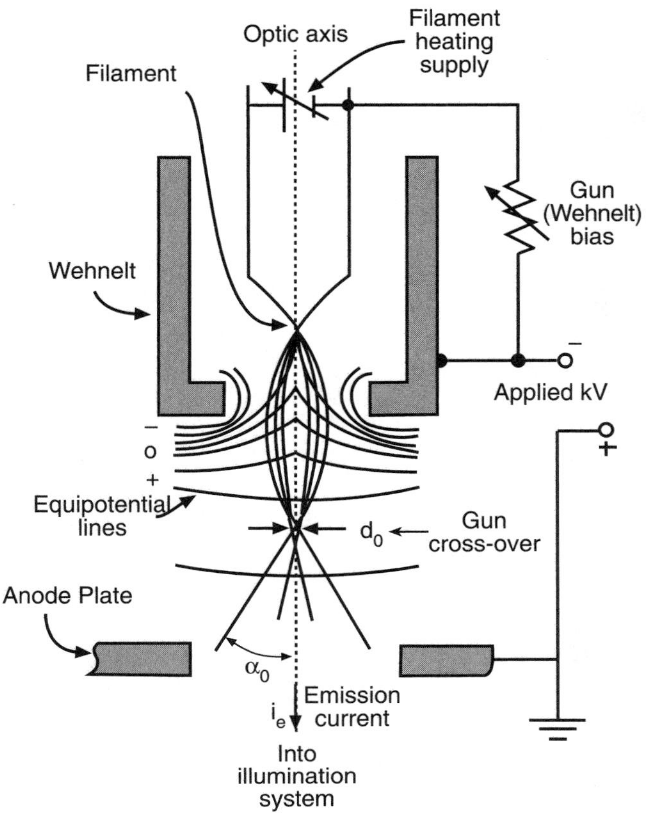

Figure 5.1. Schematic diagram of a thermionic electron gun. A high voltage is placed between the filament and the anode, modified by a potential on the Wehnelt which acts to focus the electrons into a crossover, with diameter d_0 and convergence/divergence angle α_0.

voltage power supply. This cable also connects to the tungsten filament to supply a current to heat the filament resistively to the operating temperature. LaB_6 sources are indirectly heated usually by bonding them to a metal filament such as rhenium, which is resistively heated.

> As the filament current (i_f) increases the temperature increases until thermionic emission occurs, and an emission current from the cathode i_e can be measured. Sometimes you'll find this current referred to as the "beam current," but this is misleading, because the true beam current is that which enters the specimen after the electrons have left the gun and gone through the illumination system of the microscope.

When the electrons leave the cathode they have a negative potential of 100 kV with respect to the earthed anode, so they accelerate through this potential difference acquiring an energy of 100 keV, and a velocity of greater than half the speed of light.

Now to get a controllable beam of electrons through the hole in the anode and into the microscope itself, we apply a small negative bias to the Wehnelt cylinder. The electrons coming off the cathode see the negative field and are converged to a point called a crossover between the Wehnelt and the anode as shown in Figure 5.1. We could operate the cathode heating and the Wehnelt bias controls independently, but the electronic circuitry of the gun is designed so that as the emission current increases the Wehnelt bias increases, and this arrangement is called a "self-biasing" gun. The result is shown in Figure 5.3, which plots the filament emission current (i_e) against the current used to heat the filament (i_f). As you can see, i_e reaches a maximum such that further increase in i_f doesn't increase the current going into the microscope. This is the *saturation condition* and all thermionic sources should be operated at or just below saturation. Operating above saturation reduces filament life without any compensating ad-

Figure 5.2. The three major parts of a thermionic gun, from top to bottom: the cathode, the Wehnelt cylinder, and the anode, shown separated. The Wehnelt screws onto the cathode (filament) support and both are attached to the high-tension cable which contains power supplies for heating the filament and biasing the Wehnelt. The anode sits just below the Wehnelt, in the top of the TEM column.

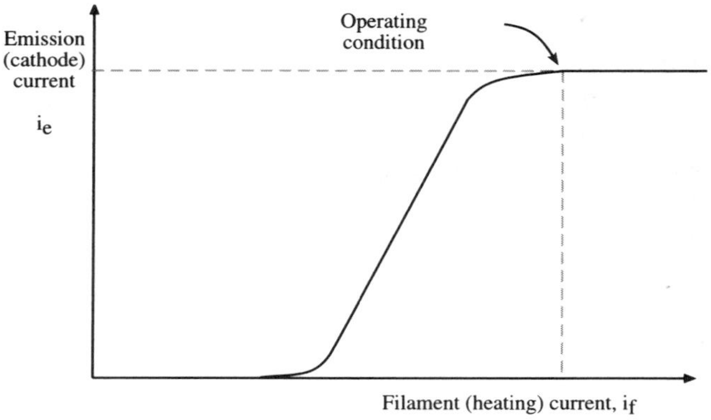

Figure 5.3. The relationship between the current emitted by the electron source (i_e) and the filament heating current (i_f) for a self-biasing gun. Increasing the filament current results in a maximum emission current termed saturation.

vantage; operating significantly below saturation reduces the current into your specimen, thus reducing the intensity of all the signals coming out of your specimen.

> **The Wehnelt acts as a simple electrostatic lens: the first lens in the microscope.**

In addition to optimizing the source life, operating at saturation also optimizes brightness. If you look at Figure 5.1, the crossover is the source size d_0 that we used

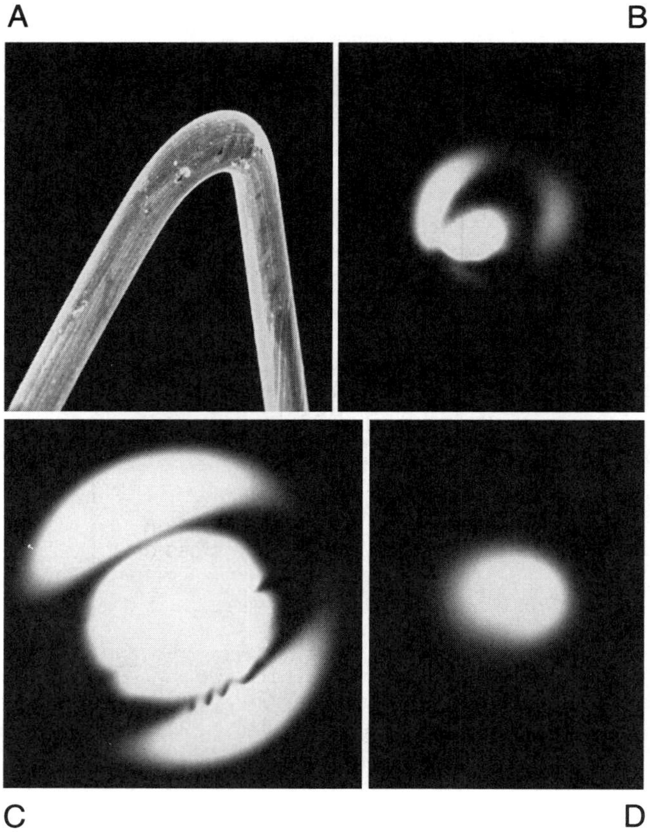

Figure 5.5. (A) The tip of a tungsten hairpin filament and the distribution of electrons when the filament is (B) undersaturated and misaligned, (C) undersaturated and aligned, and (D) saturated.

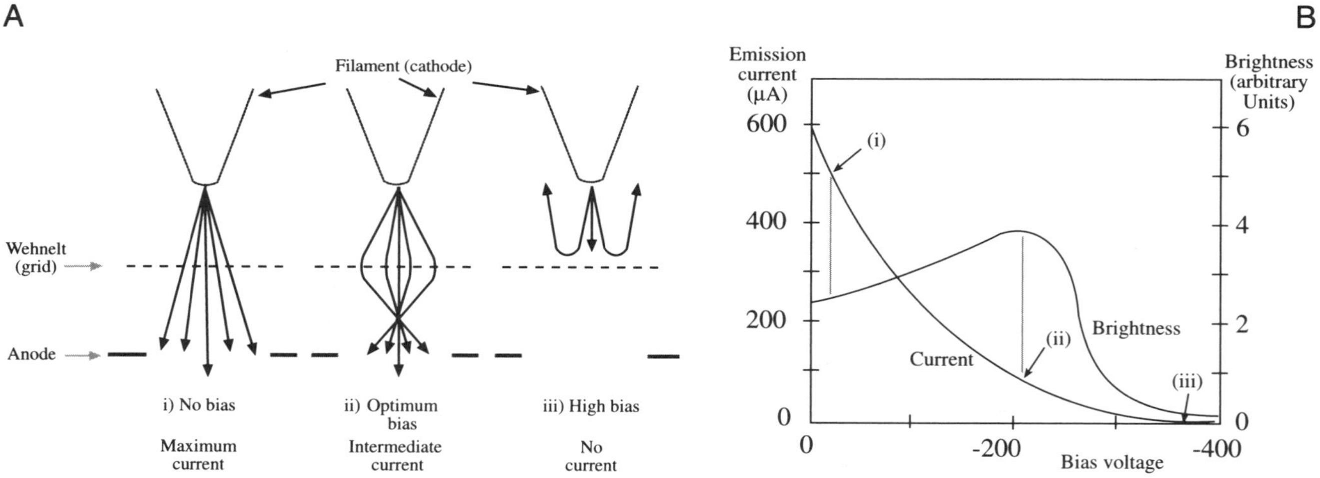

Figure 5.4. (A) The effect of increasing Wehnelt bias (i–iii) on the distribution of electrons coming through the anode. (B) The relationship between the bias and the emission current/gun brightness. Maximum brightness is achieved at an intermediate Wehnelt bias, and an intermediate emission current [condition (ii) in A].

back in the brightness equation (equation 5.3) and the convergence/divergence angle at the crossover is α_0 in that same equation. The current in the crossover is the emission current i_e. Now, as shown in Figure 5.4A, if the Wehnelt bias were too low (i) d_0 would not be very small, and if the bias were too high (iii) the cathode emission current would be suppressed. In either case β would be low. The optimum β is at an intermediate bias setting (ii), as summarized in Figure 5.4B. You might think that the small bias on the Wehnelt acts against the accelerating voltage, so the true beam voltage is the applied kV minus the Wehnelt bias (which may be up to 2 kV), but this is compensated for in the design of the gun.

So how do we achieve saturation? One way is to look at the meter which displays the i_e and watch it rise to a maximum as i_f is continuously increased. This method may not be easy because the appropriate readouts may not be available, or if they are, they may not be very sensitive. So the standard way is to look at the image of the filament crossover on the TEM screen; this image shows you the distribution of electrons coming off the filament. As thermionic emission starts the electrons come from both the central tip of the filament and a region surrounding the tip (Figure 5.5A), and so the filament image is as shown in Figure 5.5B or C, and is characteristic of an unsaturated tungsten filament. With increasing emission the halo of emission collapses in on the central bright disk, although some structure may still be visible. The filament is truly saturated when no structure is visible (Figure 5.5D).

Since LaB_6 sources have well-defined crystal facets (Figure 5.6A) they show a slightly different undersaturated image, as you'll see in Figure 5.6B, but in essence the process is identical. It is probably best to operate an LaB_6 source at conditions just below saturation, since this will extend the source life without undue loss of signal. We'll find that there are a few occasions when undersaturated operation can be useful, because the electrons in the halo are more coherent than those in the central bright region. LaB_6 crystals are more susceptible to thermal shock than tungsten, and so you should take care when heating *and cooling* an LaB_6 source. Increasing the heating current should be done slowly, with 10 to 20 seconds' pause between each setting. This is particularly critical after you've installed a new LaB_6 source.

The appearance of the image of the source, such as we show in Figures 5.5 and 5.6, can also be used to align the gun assembly so that the beam is aligned along the optic axis of the microscope. This is the only other thing you have to do to the gun apart from saturating it. The source is usually pre-aligned by the manufacturer, so alignment should be simple when it is put inside the Wehnelt. Typically, the undersaturated source image is asymmetrical as in Figure 5.5B

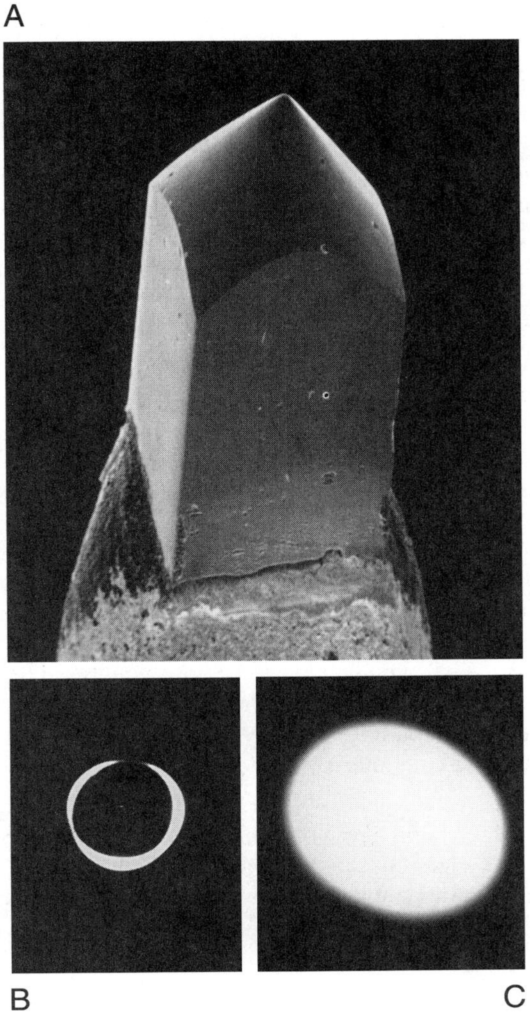

Figure 5.6. (A) An LaB_6 crystal and the electron distribution when the source is (B) undersaturated and aligned and (C) saturated.

and in those circumstances all you have to do is tilt the gun assembly to make it symmetrical as in Figure 5.5C. Detailed instructions will be in the manufacturer's handbook.

> Achieving optimum β is critical in any operations that require a fine beam (<0.1 μm).

In an SEM, which always requires a small probe, the gun is carefully adjusted by the manufacturer to produce optimum β at saturation, and you may not have any external control of the Wehnelt. In a TEM, particularly when you are operating in a broad-beam mode, there is no need to optimize β, but you may need to increase the current density and make the image appear brighter. You can do this by decreasing the Wehnelt bias, using the "emission" control. When you decrease the bias, you should go back and adjust i_f to ensure you're at saturation, since the saturation condition will change with changing bias. So now you will have a greater current density falling on the screen, but the crossover size will have increased, thus decreasing β. This is not important if you're operating with a broad beam, but if you want to operate at maximum β with a focused beam, as is the case for AEM, then you need to be able to measure β; we'll show you how to do that in Section 5.5.

5.3.B. Field-Emission Guns (FEGs)

In many ways, FEGs are much simpler than thermionic guns. In order to get an FEG to work we make it the cathode with respect to *two* anodes. The first anode is positively charged by several kV with respect to the tip. This is called the "extraction voltage" since it generates the intense electric field—extracting electrons by enabling them to tunnel out of the tip. Increasing the extraction voltage when you first switch on has to be done slowly, so the mechanical shock doesn't fracture the tip. This is the only practical step you have to carry out to run an FEG, and it can easily be computer-controlled.

- Anode 1 provides the extraction voltage to pull electrons out of the tip.
- Anode 2 accelerates the electrons to 100 kV or more.

The electrons are accelerated through the applied potential by the second anode. The combined fields of the anodes act like a more refined electrostatic lens to produce a crossover, as shown in Figure 5.7A. This lens controls the effective source size and position, but it isn't very flexible. Incorporating a magnetic lens into the gun gives a more controllable beam and larger β. The faults (known as lens aberrations) in the gun lens are very important in determining the source size; we'll talk extensively about lens aberrations in Chapter 6.

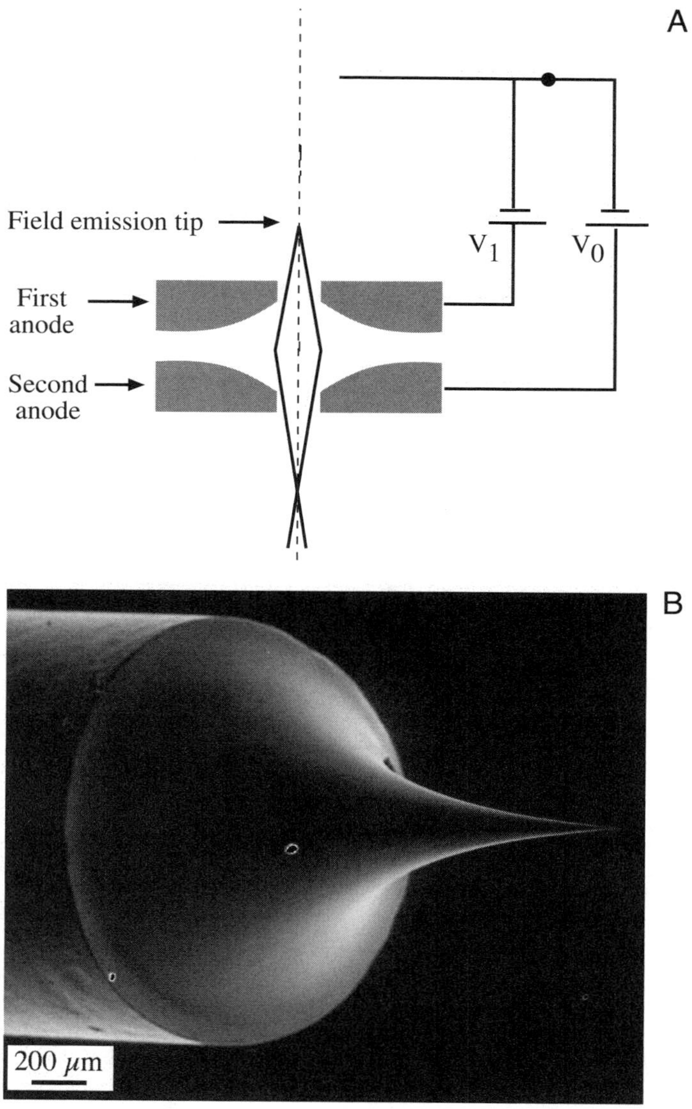

Figure 5.7. (A) Electron paths from a field-emission source showing how a fine crossover is formed by two anodes acting as an electrostatic lens. Sometimes an extra (gun) lens is added below the second anode. (B) An FEG tip, showing the extraordinarily fine W needle.

> In a vacuum of 10^{-7} Torr, one monolayer of contaminants will form on a substrate in less than a minute. At 10^{-10} Torr, it will take 7 hours to form a monolayer.

We have already noted that field emission requires a pristine surface and, even in UHV conditions, surface contaminants build up on the tip. With time, the emission current falls and the extraction voltage has to be increased to compensate. But eventually it becomes necessary to remove the contamination by "flashing" the tip. This just means reversing the potential to the tip and "blowing off" a surface layer of atoms, and/or heating the tip quickly to ~5000 K to evaporate the contaminants. In most FEGs flashing occurs automatically, when the extraction voltage increases to a certain predetermined level. Thermally assisted FEGs do not form the same surface contamination layer and so don't need flashing. A typical FEG tip is shown in Figure 5.7B.

5.4. COMPARISON OF GUNS

All the important characteristics of the three guns we've talked about are summarized in Table 5.1. Tungsten sources are the worst in most respects, but for routine TEM applications they are excellent, reliable sources and are cheap, robust, and easily replaceable.

LaB_6 is a more useful source for several reasons. While it is not as refractory as tungsten, LaB_6 has a much lower value of Φ, and since Φ appears in the exponential in the Richardson equation, its effect on the current density is dominant. LaB_6 crystals can be produced with a fine tip about 1 μm in radius, which accounts for the smaller crossover size. As a result LaB_6 current densities are considerably higher than for tungsten. The brightness is typically 10 times that of tungsten, even though LaB_6 is usually operated at a much lower T to increase operating life. The decreased source size also results in improved coherency and the energy spread can be as little as 1 eV.

The drawback to LaB_6 is purely economic. LaB_6 sources cost several hundred dollars each while tungsten filaments are so cheap that the manufacturer often provides them free. Because LaB_6 is a highly reactive material, the gun vacuum has to be 10–100 times better than for tungsten, and is correspondingly more expensive to construct. So if the cost is not the criterion, LaB_6 guns are *the* recommended thermionic source, for all aspects of TEM, but particularly AEM. The increased brightness, higher coherency, and longer life are tremendous advantages. But you as the operator have the most control over its performance and you can most easily destroy it by careless heating and cooling and oversaturation. So treat LaB_6 sources gently and you will be well rewarded. If users are not careful, your TEM supervisor may try to extend the life of the LaB_6 to the point where it behaves no better than a W filament. LaB_6 sources don't die, they fade away.

In FEGs, the current density is enormous and β is correspondingly high. The values in Table 5.1 are all for 100-kV accelerating voltage and you should remember that for the tungsten and LaB_6 sources, β increases linearly with kV, so there are advantages to using 300 and 400 kV instruments, although the thermionic source brightness at 400 kV still does not approach β of an FEG at 100 kV. The extremely small source size means that the beam is highly spatially coherent and the resulting energy spread is minuscule for cold FEGs; thermally assisted FEGs give a larger energy spread. So for all applications that require a bright, coherent source, the FEG is best. This is the case for AEM, HRTEM, and such special applications as electron holography and Lorentz microscopy (for looking at magnetic domains). However, as we'll see later, the coherence of the source may produce a new complication: we must interpret the image!

For routine TEM, an FEG is far from ideal because the source size is so small. It is thus not possible to illuminate large areas of the specimen without losing current density, and therefore intensity, on the screen. Under these circumstances, a thermionic source is better. This limitation to FEG applications may be overcome by the larger *p-n* FE sources, which use small (≈ 1–10 μm) Si semiconductor crystals, but this is still a new and developing technology.

Another drawback to FEGs is the need for UHV conditions. UHV technology is expensive and requires a much higher level of operator competence. As a result, FEG TEMs are relatively rare. But in the SEM field there is a whole new generation of computer-controlled low-voltage instruments, and it will only be a matter of time before FEG TEMs are common.

TABLE 5.1. Characteristics of the Three Principal Sources Operating at 100 kV

	Units	Tungsten	LaB_6	Field Emission
Work function, Φ	eV	4.5	2.4	4.5
Richardson's constant	A/m^2K^2	6×10^5	4×10^5	
Operating temperature	K	2700	1700	300
Current density	A/m^2	5×10^4	10^6	10^{10}
Crossover size	μm	50	10	<0.01
Brightness	A/m^2sr	10^9	5×10^{10}	10^{13}
Energy spread	eV	3	1.5	0.3
Emission current stability	%/hr	<1	<1	5
Vacuum	Pa	10^{-2}	10^{-4}	10^{-8}
Lifetime	hr	100	500	>1000

5.5. MEASURING YOUR GUN CHARACTERISTICS

This section requires that you know how to operate a TEM. If you're a novice, you should skip this part of the chapter for now because we are going to refer ahead in the book for much of what you need to know.

For conventional TEM imaging and diffraction and many other routine uses, all you need to do is saturate and align the gun and then ignore it. There are, however, times when we need to be able to measure the brightness and coherency. The source brightness is a most important parameter to measure in an AEM since, if the gun is not operating at its maximum β, then the quality of the analytical information that is generated will be poor. Similarly, knowing the energy spread of your source is important for electron spectroscopy, and having a measure of the beam coherency can be important for some more advanced techniques that we've just mentioned. So let's see how we can measure the various parameters that we've just discussed. We'll start with β, then ΔE, and finally the coherency.

By measuring the three variables in equation 5.3, i.e., the beam current, the beam diameter, and the semiangle of convergence, we can determine β. However, while we can easily get a measure of the emission current at the gun, it is more difficult to measure d_0 and α_0 there. So we make the approximation that, if we neglect lens aberrations, β is constant throughout the electron optical system so it doesn't matter where it is measured. It is easiest, practically, to determine β at the plane of the specimen and we'll now show you how to do this.

5.5.A. Beam Current

You can measure the beam current at the specimen i_b directly using a Faraday cup in a specimen holder. A Faraday cup consists of a small aperture above a relatively deep hole in an earthed metal block. If the aperture is small enough (e.g., about 50 μm) and the metal block deep enough (about 2 mm), and made of something light like Al to minimize backscatter, then it is a reasonable assumption that no electrons escape back out of the entrance aperture. All the electrons going into the aperture therefore go to earth, and you can measure the electron current using a picoammeter in the earth line. Ideally a Faraday cup should be available permanently in the column of a TEM, and this would permit constant monitoring of the beam current. You can also calibrate the Faraday cup measurement against the TEM screen exposure meter or the electron energy-loss spectrometer shield current. This procedure permits you to make a more rapid estimate of i_b at any time you need it.

As we'll show in Chapter 9, i_b is a strong function of the beam size. Therefore the current is controlled by the first condenser (C1) lens strength, and the size of the final beam-limiting aperture in the second condenser (C2) lens. If you look ahead to Figures 9.10 and 9.11 you will see the variation of i_b as a function of C1 lens strength and the effect of C2 aperture size on α.

- The beam current is usually in the range from nanoamps to picoamps.
- The emission current is typically several microamps.

Most of the current from the gun is lost in the illumination system, as we'll see in Chapter 9.

5.5.B. Convergence Angle

You can easily measure the convergence semiangle α from the convergent-beam diffraction pattern, which you can see directly on the TEM screen. (You will need to read Chapter 21 on convergent-beam diffraction in order to find out how to generate such patterns.) In the schematic diagram in Figure 5.8, the total convergence angle 2α is proportional to the width of the diffraction disks, *a*. This width can easily be calibrated if the specimen has a known Bragg angle $2\theta_B$ (see Chapter 11), since $2\theta_B$ is proportional to the distance, *b*, from the 000 disk to the *hkℓ* disk. Thus

$$2\alpha = 2\theta_B \frac{a}{b} \qquad [5.6]$$

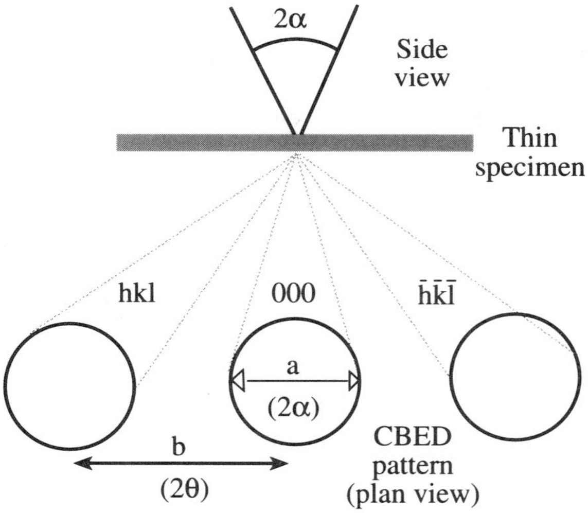

Figure 5.8. The distances on a convergent-beam diffraction pattern from which you can measure the beam-convergence semiangle, α, which is proportional to the width of the diffraction disk.

The convergence semiangle is not only important in the brightness equation, but we'll see that it also plays a major role in convergent-beam patterns, in STEM imaging, and in EELS. Knowledge of α is useful in many aspects of TEM. The value of α is controlled by the size of the final limiting aperture in the illumination system and we'll see how this works in Chapter 6.

5.5.C. Calculating the Beam Diameter

While it is a relatively simple matter to measure i_b and determine α, the measurement of d, the beam diameter, is not so straightforward. However, d is a major factor in all aspects of TEM where we use a fine focused beam, such as AEM and STEM imaging. We can either calculate d or measure it experimentally.

The first problem with determining d is that there is no universally accepted definition of the beam diameter. The manufacturer will give you a list of nominal beam sizes for each setting of the C1 lens. These values are *calculated* and may differ from the actual beam size by large amounts. The calculation assumes that the electron intensity distribution in the beam is Gaussian, and the beam diameter is defined as the full width at half maximum (FWHM) of the Gaussian distribution, defined in Figure 5.9. To approach a Gaussian intensity distribution, the beam must be well aligned, any astigmatism in the condenser lenses corrected (see Chapter 9), and all apertures in the illumination system accurately centered. Even under these conditions you cannot obtain Gaussian conditions for every possible beam size. For example, there may be six different C1 lens excitations, each of which gives a different calculated beam size, but there are invariably fewer than six C2 apertures available, so each beam size cannot be correctly apertured; spherical aberration effects will then broaden the beam size beyond a true Gaussian. If you select too small an aperture, then the intensity distribution will be truncated at a fraction of the full Gaussian curve.

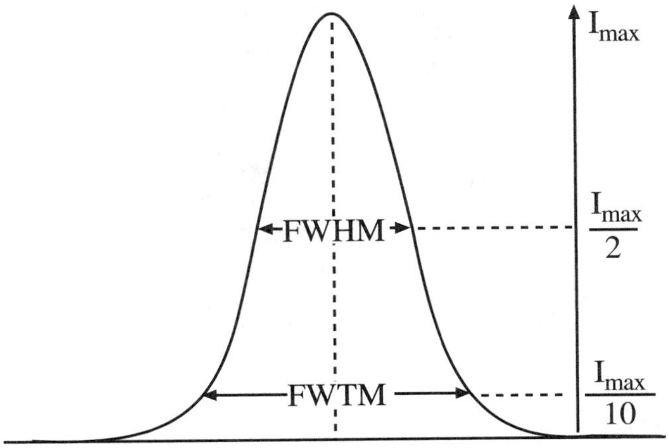

Figure 5.9. The definition of the full width at half maximum (FWHM) and the full width at tenth maximum (FWTM) of a Gaussian intensity distribution which is typical of a well-aligned beam.

To make a complete calculation of the beam size, we assume that it is determined by an initial Gaussian diameter at the gun (d_g). This diameter is broadened by the effects of spherical aberration in the beam-forming lens (d_s) and diffraction at the final aperture (d_d). All these terms should be added in quadrature to give a total, calculated beam size, d_t

$$d_t = \left(d_g^2 + d_s^2 + d_d^2\right)^{1/2} \qquad [5.7]$$

This equation gives us only a first-order estimate, since it is not clear that all the contributions are Gaussian. We'll now briefly discuss the origin of each of these terms.

The value of d_g is a function of β, and a value of β has to be assumed for the purposes of calculation. The expression for d_g is

$$d_g = \frac{2}{\pi}\left(\frac{i}{\beta}\right)^2 \frac{1}{\alpha} \qquad [5.8]$$

We have already defined i, β, and α.

The disk of minimum confusion caused by spherical aberration has a diameter given by

$$d_s = 0.5\, C_s\, \alpha^3 \qquad [5.9]$$

where C_s is the spherical aberration coefficient, which we discuss in detail in Chapter 6. This is the full diameter containing 100% of the beam current. Clearly, this term is not Gaussian unless the beam is correctly apertured which, as we just discussed, is not always possible. The calculated diameter due to diffraction is

$$d_d = 1.22\,\frac{\lambda}{\alpha} \qquad [5.10]$$

which is the Rayleigh criterion which we discussed in Chapter 1. Although all these definitions do not define the same diameter of the electron distribution, they are all combined to give a first approximation of the FWHM of the beam. Clearly, it is more reliable, but more time-consuming, to measure d experimentally. Figure 5.10 shows the result of calculations of the three contributions to the beam diameter in a VG HB501 STEM.

5.5.D. Measuring the Beam Diameter

To measure the beam size in a TEM/STEM, you must form an image of the beam on the TEM viewing screen under conditions where you know, or can calibrate, the magnification. This is a nontrivial exercise and you may need to

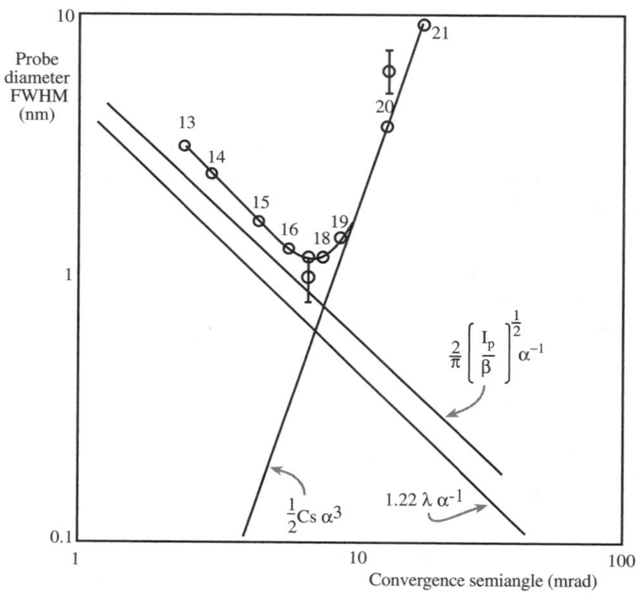

Figure 5.10. Calculations of the three contributions to the probe size as a function of the convergence semiangle α in an FEG STEM with a probe current I_p of 0.85 × 10^{-8} A. Two experimental measurements are shown, at condenser 1 lens settings 17 and 20. The minimum probe dimension is ~1 nm with α < 10 mrads.

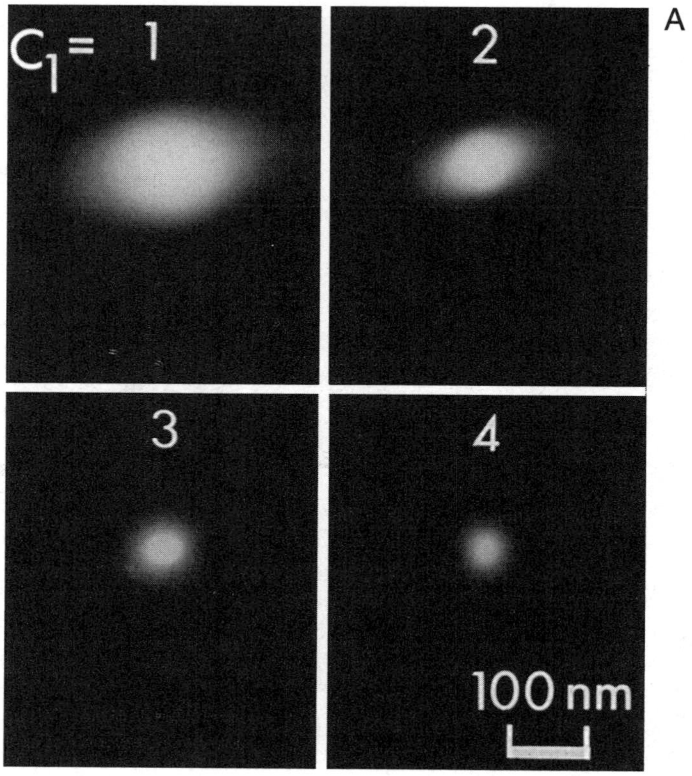

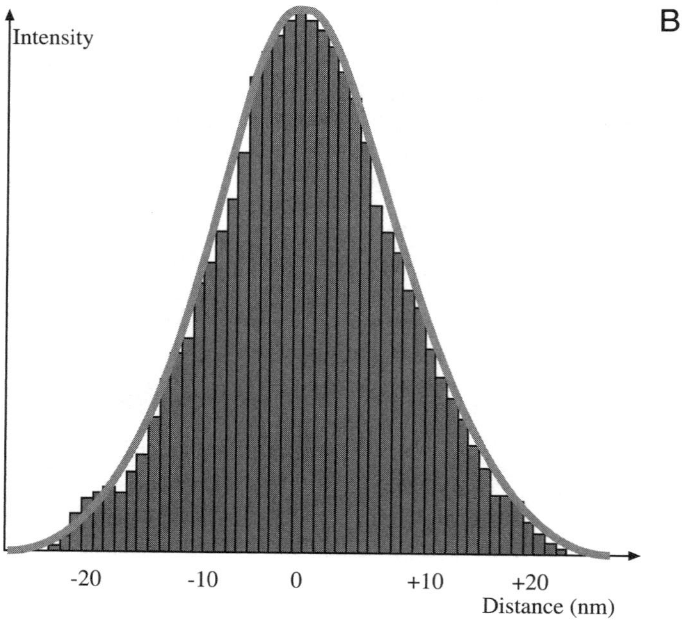

Figure 5.11. (A) Four images of the beam formed on the TEM screen at different condenser 1 lens settings and (B) the corresponding microdensitometer trace across spot #3, confirming the Gaussian nature of the intensity distribution.

consult the manufacturer to be sure that you are doing it correctly. You can then photograph the beam and determine the intensity distribution from a microdensitometer trace across the image, as shown in Figure 5.11.

- The FWHM contains 50% of the integrated intensity. It is the value used by the manufacturers when they report beam sizes. It is also the important dimension when considering the effect of *d* on the STEM image resolution.
- The full width at tenth maximum (FWTM) contains 90% of the integrated intensity. It is a more relevant dimension because the Faraday cup measures the current in the total beam which is closer in size to the FWTM.

When you insert the beam diameter in the brightness equation, either the FWHM or the FWTM can be used. The FWTM is equal to 1.82 x FWHM and this is also shown in Figure 5.9. You should note, therefore, that you overestimate β if you use the smaller FWHM. Use of the FWTM is also the preferred beam size when calculating the spatial resolution of microanalysis, as we describe in Chapter 36.

In a dedicated STEM you can't image the beam directly, since there are no post-specimen lenses to magnify

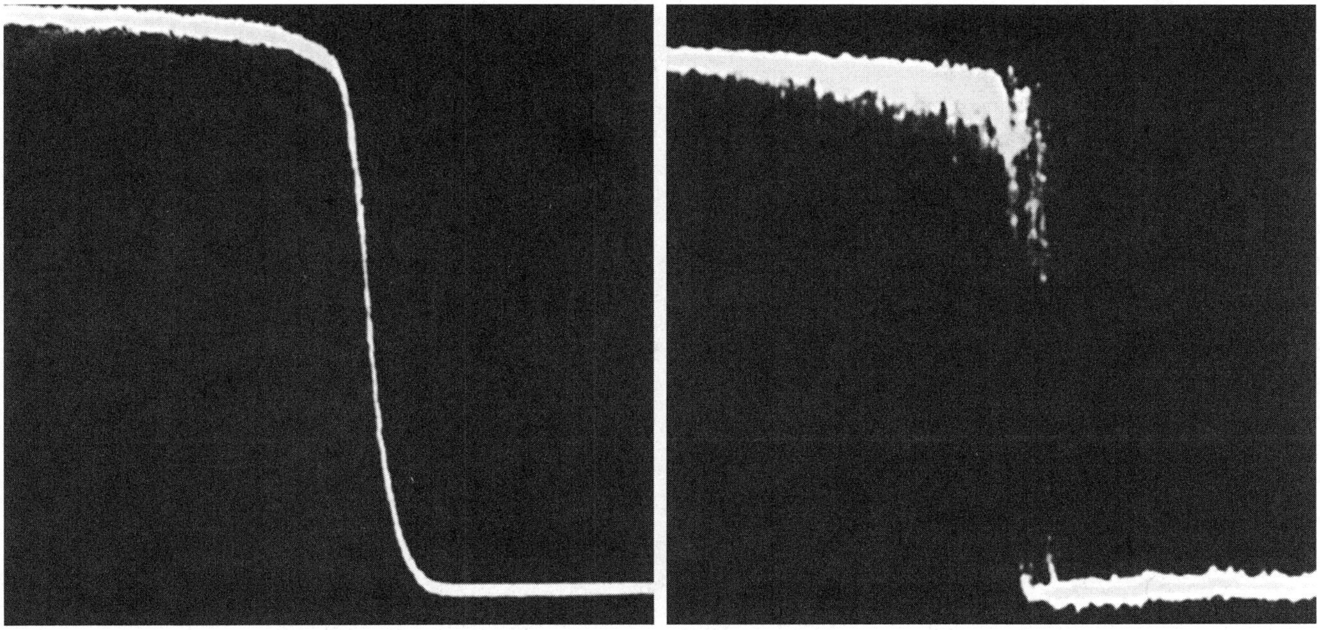

Figure 5.12. Intensity profiles obtained by scanning a fine beam across a sharp edge of a cube of MgO. The measured probe size (FWTM) in (left) is 7.4 nm (magnification 1×10^6) and in (right) 1.8 nm (magnification 11×10^6). The smaller probe contains a much smaller current and is therefore a noisier trace.

A B

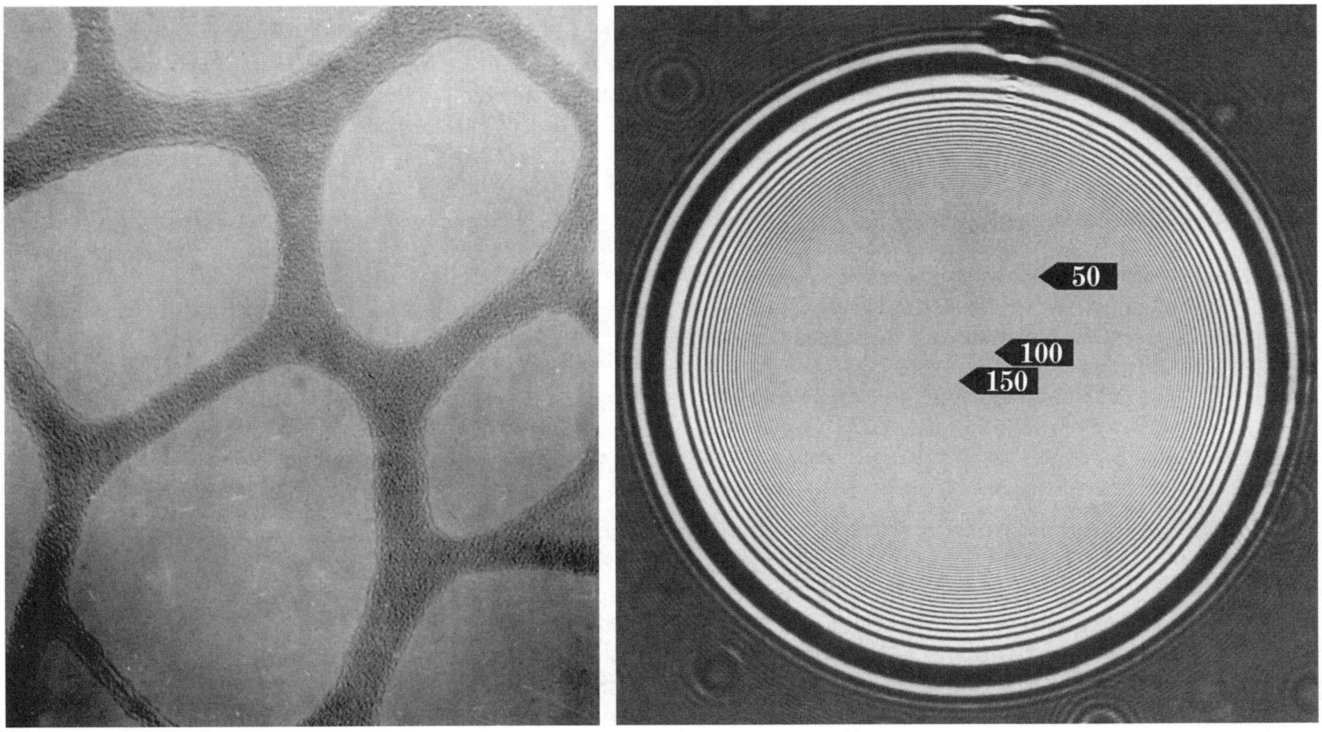

Figure 5.13. Fresnel fringes from (A) a thermionic source with poor coherency and (B) an FEG with high coherency.

its image and no photographic film to record it. The value of d must be determined indirectly, as in other scanning instruments. The best method involves scanning the beam across a knife-edge specimen and monitoring the intensity change which occurs, for example, by recording the output from the annular dark-field detector. This approach yields an integrated intensity profile, as shown in Figure 5.12. In order to extract a value of the FWHM or FWTM from the profile, you must make measurements between various points determined by integrating the intensity from one side of a two-dimensional Gaussian to the other. In Figure 5.10, two experimental beam-size measurements are shown; they show reasonable agreement with the calculated values from the brightness equation.

The measurement of d is clearly not a simple procedure. You can find a full description of the problems in the paper by Michael and Williams (1987).

5.5.E. Energy Spread

Remember that the energy spread (ΔE) of the electron beam is a measure of the temporal coherency. This spread is important in EELS and, in fact, the only way to measure the energy spread is to use an electron spectrometer. Under conditions where the spectrometer itself is not limiting the resolution of the spectrum, the value of ΔE can be simply measured by collecting a spectrum of electrons without a specimen in the way of the beam. The spectrum then consists of a single Gaussian peak and the resolution of the spectrum is defined as the FWHM of this peak. You can find out how to do this in detail in Chapter 37. Typical values of ΔE for the various electron sources are also given in Table 5.1.

5.5.F. Spatial Coherency

It's difficult to measure the coherency of the beam experimentally although, as we've discussed, small sources ensure spatial coherency. One practical way of measuring the coherency is to form an image of the edge of a hole in a specimen, such as a thin holey carbon film. When you operate slightly out of focus you see alternating dark and bright fringes, called Fresnel fringes, as shown in Figure 5.13A. Typically for a thermionic source only one or two fringes are visible. These fringes are a phase-contrast effect (see Part III). We can also use them to correct the astigmatism in the objective lens, as we'll see in Chapter 9. The number of visible fringes is a measure of the beam coherency. Figure 5.13B shows the enormous number generated by an FEG.

5.6. WHAT kV SHOULD YOU USE?

For the materials scientist, this is usually an easy question to answer. You always operate at the maximum available kV, unless there is a definite reason to use a lower kV. Of these reasons, the most obvious is avoiding beam damage, but we'll see others later in the book, so don't forget that you can always operate a 300-kV machine at 100 kV. Remember, it's like being able to change the wavelength of a monochromatic light source in a visible-light microscope (VLM). The threshold for beam damage for most metals is less than 400 kV, which is the highest available voltage on "off-the-shelf" TEMs. For lighter and more beam-sensitive materials, such as some ceramics and polymers, lower voltages may be better, but there is not much use going below 100 kV since the images will be rather dim and you'll have to make extraordinarily thin specimens to see anything useful. The reasons for choosing the highest kV are:

- The gun is brightest.
- The wavelength is shortest; the resolution is potentially better.
- The cross section for inelastic scatter is smaller; the heating effect is smaller.

CHAPTER SUMMARY

Most TEMs use thermionic sources and, if you have the choice, use an LaB_6 source and run at the highest kV. Take care when heating and cooling the LaB_6 crystal and always operate just below saturation to maximize the lifetime of a source. If you're going to be doing AEM, get some idea of the beam current that you can get from your source under typical operating conditions. Also, measure the beam size and convergence angle to give a measure of β, and if you're doing EELS then the energy spread is essential information. If you have an FEG you'll most likely be doing fine probe analytical work, in which case all the above characteristics must be measured, and if you're going to do high-resolution imaging, then the degree of coherency is important too. Always treat the source carefully when changing it, aligning it, saturating it, or switching it off. There's nothing more annoying than losing your source, since it usually happens at some critical point during your work.

REFERENCES

General References

Broers, A.N. (1974) SEM 1974 (Ed. O. Johari), p. 9, IITRI Chicago.
Joy, D.C. (1973) SEM 1973 (Ed. O. Johari), p. 743, IITRI Chicago.
Orloff, J. (1989) *Ultramicroscopy* **58**, 88.
Reimer, L. (1993) *Transmission Electron Microscopy; Physics of Image Formation and Microanalysis*, 3rd edition, Springer-Verlag, New York.
Veneklasen, L.H. (1975) *Optik* **36**, 410.

Specific References

Hawkes, P.W. (1978) *Advances in Optical and Electron Microscopy* **7**, 101.
Michael, J.R. and Williams, D.B. (1987) *J. Microsc.* **147**, 289.

Lenses, Apertures, and Resolution

6

CHAPTER PREVIEW

Electron lenses are the magnetic equivalent of the glass lenses in an optical microscope and, to a large extent, we can draw comparisons between the two. For example, the behavior of all the lenses in a TEM can be approximated to the action of a convex (converging) glass lens on monochromatic light. The lens is basically used to do two things:

- either take all the rays emanating from a point in an object and recreate a point in an image,
- or focus parallel rays to a point in the focal plane of the lens.

The lens can't collect *all* the rays from the object and we often deliberately limit the collection angle with an aperture. We can draw ray diagrams showing how electron lenses control beams of electrons. These diagrams correspond directly to the ray diagrams used in physical optics. Of course the analogy with light fails for certain aspects, but basically it will pervade this chapter. So we'll start by reminding you of the principles of light optics insofar as they relate to electron optics. Then we'll discuss the magnetic electron lens in more detail, showing how an electron behaves as it passes through such a lens. We'll describe some actual lenses and tell you how we use different kinds of electron lenses to do different things in the microscope.

The major limit to the use of electron lenses is the fact that we aren't very good at making them. They suffer from rather severe aberrations, which we control by inserting limiting apertures. You need to understand these aberrations, since they play a major role in deciding what we can and cannot do with the microscope. In particular, the resolution of an electron lens (rather than the wavelength of the electrons) limits the resolution of the TEM. Since resolution is usually the single most important reason for buying a TEM, you need a firm understanding of this concept. Unfortunately, we electron microscopists aren't very firm in our definitions of resolution. Finally, we describe how the apertures we put in the lenses aid both the depth of field and the depth of focus of the instrument.

Lenses, Apertures, and Resolution

6

6.1. WHY LEARN ABOUT LENSES?

Why should we learn about electron lenses? As in a visible-light microscope, the lenses in a TEM control all the basic operational functions of the instrument. We physically move glass lenses up and down in a light microscope to control the intensity of the illumination and the focus and magnification of the image. The focal length of a glass lens is fixed. In a TEM the positions of the lenses are fixed and we focus, etc., by changing the strength of the lenses. As you'll see, in most cases the lenses we use are magnetic, so that we change their strength by changing the magnetic field. Almost any operation we carry out on the TEM involves changing magnification or focus; we use electron lenses to magnify and focus the electron beam, the images, and the diffraction patterns.

> We also use apertures in the lenses to control the beam current and the convergence of the beam hitting the specimen.

These factors are critical in imaging, diffraction, and microanalysis. An aperture is used to select different electron beams to form different images, thus manipulating the image contrast. Another aperture is used to select different regions of the specimen to contribute to the diffraction pattern.

In essence then, we control the quality of our images, diffraction patterns, and analytical signals by adjusting the lenses and their apertures. So knowing how these aperture/lens combinations work allows you to understand how we control the TEM and why we do certain operations on the microscope.

An understanding of electron lenses will help us to answer such questions as:

- Why can we see finer detail with an electron microscope than with a light microscope?
- Why can't we see as much detail as we might expect from physics?
- Why does the TEM have a better depth of field and depth of focus than the light microscope?

We'll see that the answer to these questions lies in the quality of the lenses, and how we use them. In this chapter we'll discuss the basics of how a lens/aperture combination works. Throughout the book you'll come across different uses and combinations of lenses and apertures. So this is a central chapter for the serious microscope operator.

6.2. LIGHT OPTICS AND ELECTRON OPTICS

You are already familiar with the action of a magnifying glass lens on light rays. The magnifying glass is a convex lens. It can be used in two ways to control the light rays coming through it. First, it can produce a magnified image of the object you're looking at. Second, it can focus a parallel beam of light to a point, in the focal plane of the lens. (As children, we used this latter property to set fire to a piece of paper by focusing the sun's rays.) These two actions, forming an image of an object and focusing parallel rays to a point, are all we need in order to understand how the lenses in a TEM work. The reason that we can get away with this simple approach is because the electron lenses that we use act, to a reasonable approximation, like convex glass lenses; in detail, they're often equivalent to more complex combinations of convex lenses. Remember that, at present, all magnetic lenses are convex lenses.

6.2.A. How to Draw a Ray Diagram

In traditional light optics it's customary to draw ray diagrams of the path of light rays through the lens, and we do the same for electrons and their lenses. These ray diagrams are usually drawn horizontally because the traditional optical bench on which light optics experiments are carried out is a horizontal setup. But since the electron microscope is usually a vertical instrument, we will draw all our ray diagrams vertically.

Let's start by drawing ray diagrams to illustrate the two fundamental actions of image formation and focus of parallel rays. In these and all subsequent diagrams we'll draw all the lenses in the TEM as convex lenses. We will draw all ray paths as straight lines outside the lens, and we'll start by assuming that the lenses are perfect. We'll also draw the lenses as so-called "thin" lenses, which means their thickness is small compared to their radii of curvature. Actually, we'll make the lenses *very* thin. We'll see later that all these assumptions are wrong, to a degree, but that these traditional illustrations are nonetheless useful.

The first thing we need to do is to have a base line on which to draw our diagrams; this line is called the optic axis.

> The optic axis is an imaginary line down the column of the TEM passing through the center of each lens.

Now the first action of a lens that we want to show is how it produces an image of an object. In a TEM the object will usually be the specimen itself or an image of it, but it may also be the electron source, which is an object for the illumination system. If we assume the object is a point and the radiation is emanating from that point (a so-called "self-luminous object"), then a perfect lens will gather a fraction of that radiation and form a point image. This action is shown in Figure 6.1 in which the point is on the optic axis. The fraction of the rays from the object gathered by the lens is an important variable, defined by the semiangle β in Figure 6.1. Ultimately, as you can see, β is governed by the size of the lens, but we often choose to limit β by inserting an aperture, as we'll discuss later in this chapter. You'll often see the semiangle of collection defined as α, but we will reserve α for convergence semiangles (see Section 2.7).

> So, all lenses are imperfect insofar as they cannot gather all the radiation emitted by an object and so we can never create a perfect image.

As you know from Chapters 2–4, most electrons are strongly forward scattered, so we can in practice gather a high fraction of the scattered electrons.

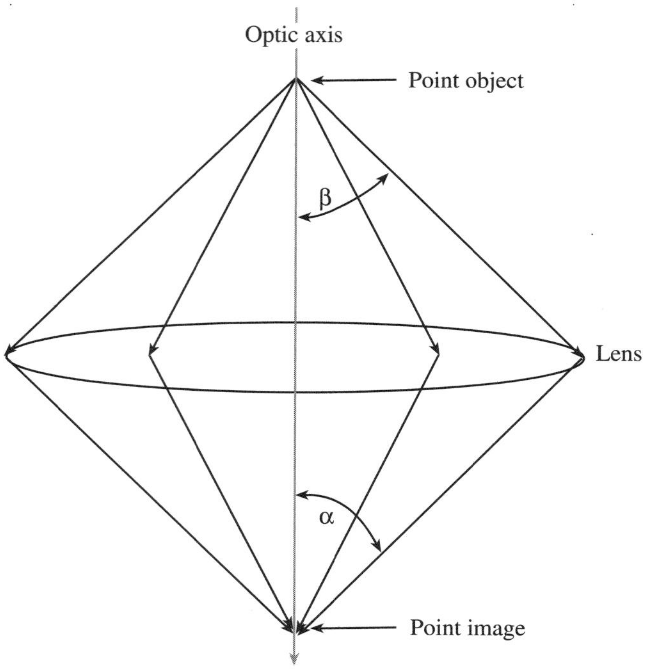

Figure 6.1. Image formation by a convex lens. A point object is imaged as a point and the collection semiangle of the lens is defined relative to the object (β) or the image (α).

> The angles in Figure 6.1 and in the other ray diagrams we'll draw are all greatly exaggerated. In practice, a typical value of β is maybe a few tens of milliradians (10 mrads = 0.57°) so if the diagrams were drawn to scale they would be many times longer than they were wide and all the ray paths would be exceedingly narrow. Since drawing to scale is impractical we always exaggerate the angles considerably in all electron ray diagrams.

If the object has a finite size, we can illustrate this by an arrow, asymmetrically positioned with respect to the optic axis, as in Figure 6.2. Then the lens creates an image of the arrow, rotated by 180°. To draw this figure, the first step is to draw line 1 from the arrowhead through the center of the lens, because rays crossing the optic axis in the lens (or "on-axis" rays which travel down the axis) are *not* affected by the lens at all and remain as a straight line. (Of course, this is a fundamental principle of how a lens works.) The second step is to draw line 2, which is a ray from the arrowhead that is parallel to the optic axis. The farther away that rays are from the optic axis, the more strongly they are bent by a convex lens, so we take line 2 and bend it toward the optic axis as it passes through the

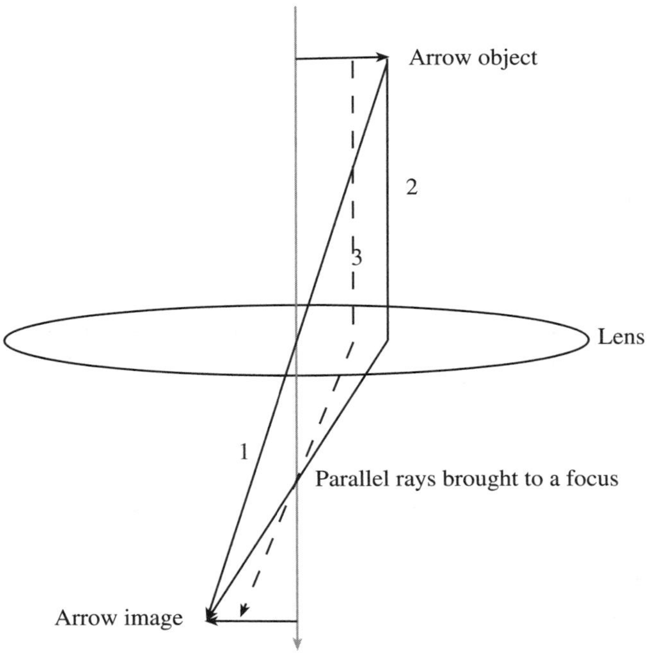

Figure 6.2. How to draw a ray diagram: first construct ray 1 through the middle of the lens, then ray 2, parallel to the optic axis, to determine the lens strength. Finally, draw line 3 parallel to 2 to define the focal plane where the parallel rays are focused. Thus an asymmetric object is imaged off axis and rotated through 180°.

lens. We can choose to make the lens as strong as we wish, and the strength determines how much the ray is bent and where lines 1 and 2 meet to recreate an image of the arrowhead. We could draw as many rays as we wished from any point on the object arrow to an equivalent point on the image arrow, such as line 3. Note that parallel rays 2 and 3 both cross the axis at the same point, illustrating the second fundamental action of a convex lens, i.e., bringing parallel rays to a focus. Again, the strength of the lens determines where the parallel electrons are focused.

> The image formed after each lens is rotated by 180° with respect to the object.

Now a full ray diagram for an object of finite size, symmetrically positioned about the axis, combines aspects of Figures 6.1 and 6.2, as shown in Figure 6.3. In Figure 6.3, all rays from a point in the object are brought back to a point in the image and all parallel rays (whether parallel to the optic axis or not) are brought to a focus in a plane at a position depending on their angle to the axis. Note that on-axis parallel rays are focused on axis and off-axis parallel rays are focused off axis. This is a most important property, since it allows the lens to create diffraction patterns in their focal plane. We'll use this diagram to introduce you to the principal terms used in lens optics.

6.2.B. The Principal Optical Elements

From the above diagrams, we can define three important planes to which we will often refer. The first plane is the object plane, which is the plane containing the object point in Figure 6.1 or the object arrow in Figures 6.2 and 6.3. The object plane always lies above the lens in question in the diagrams in this text. The second plane is the image plane (sometimes called the Gaussian image plane), which is the plane containing the image point or arrow, and it always lies below the lens. These two planes are said to be "conjugate," which means "optically equivalent." Rays leaving a point in one plane are brought to a point (if the lens is perfect) in a conjugate plane and vice versa. In other words, the electron doesn't care which way it goes through the lens. The third plane is the focal plane of the lens, and this is the plane in which the parallel rays are brought to a focus as shown in Figures 6.2 and 6.3. In the image-forming process in a TEM, the focal plane lies after or "behind"

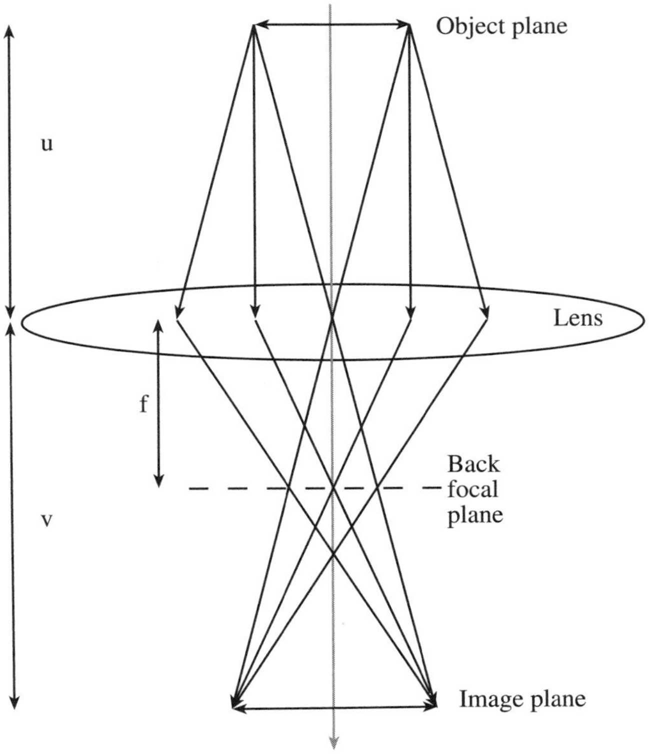

Figure 6.3. A complete ray diagram for a finite object, symmetrically positioned around the optic axis. All rays emerging from a point in the object (distance u from the lens) that are gathered by the lens converge to a point in the image (distance v from the lens) and all parallel rays are focused in the focal plane (distance f from the lens).

the lens and so the plane is sometimes called the "back focal plane" (BFP). There is also an equivalent "front focal plane," and a convex lens would take all the rays coming from a point in the front focal plane and create a parallel beam of radiation, in exactly the reverse manner to Figures 6.2 and 6.3.

6.2.C. The Lens Equation

From the above diagrams we can define three important distances, labeled in Figure 6.3: the distance from the object plane to the lens (the object distance, u), the distance from the lens to the image plane (the image distance, v), and the distance from the lens to the back focal plane (the focal length, f). Now if the lens is symmetric in strength either side of the lens plane (i.e., the front and back focal planes are the same distance from the lens), then we can write the following basic equation

$$\frac{1}{u} + \frac{1}{v} = \frac{1}{f} \qquad [6.1]$$

which is known as Newton's lens equation, and its proof can be found in a standard optics text such as Jenkins and White (1976). In thick lenses u and v are measured from different principal planes in the lens, but from the same plane in the middle of a thin lens, which we are assuming here. In all cases that we'll consider, the object distance (and therefore the image distance) is greater than the focal length. Thus a real image is produced on the other side of the lens beyond the back focal plane. If the object were within the (front) focal length, then a virtual image would be produced on the same side of the lens as the object, and this is often the case in light optics. Since we don't deal with virtual images in the TEM we'll ignore this aspect.

6.2.D. Magnification, Demagnification, and Focus

We can use Newton's lens equation to define the magnification of the convex lens as

$$M = \frac{v}{u} \qquad [6.2]$$

M is also approximately equal to the ratio of the collection semiangles of the lens subtended at the object (β) and at the image (α) as shown in Figure 6.1, assuming that these angles are small, as they invariably are in a TEM. In this example the magnification is unity.

Now we may sometimes want to *demagnify* an object (for example, when we want to form a small image of the electron source, to create the finest possible beam at the specimen). If that is the case, we define the demagnification as $1/M$. In an optical microscope we could change the magnification by moving the object relative to the lens or vice versa, and adjusting our eyes accordingly, but generally we rotate in another objective lens of different strength (curvature). In an electron microscope, we change magnification in this latter way by changing the strength of the lens, but you'll see that we can do this without changing the lens itself. So electron lenses differ fundamentally from glass lenses in that one lens can be adjusted to a range of strengths.

If we make the lens stronger, then the focal length is shortened as shown in Figure 6.4. If f is shortened but u is unchanged, then v must be correspondingly shorter and the image magnification is smaller, or the demagnification is larger. Under these conditions which normally occur in the TEM, strong lenses magnify less and demagnify more, which is counter to our understanding of light microscopes in which stronger lenses produce greater magnifications.

How do we get the high magnification that we need to form images of atom rows such as Figure 1.2? What we do is put the object close to the lens, making u small and M large, and repeat this for several lenses in tandem one after the other. So we end up with a multilens system like a com-

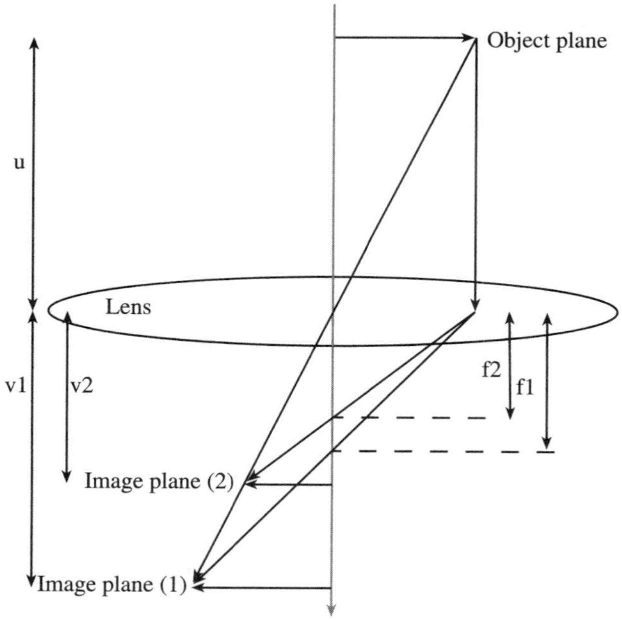

Figure 6.4. Strengthening the lens shortens the focal length f. So a weaker lens ($f1$) produces a higher magnification of the object than a stronger lens ($f2$) since the image distance v increases, but the object distance is unchanged.

pound optical microscope. (We'll discuss the details of the lens combinations in the illumination and imaging systems of the TEM in Chapter 9.) In these circumstances, the image plane of the upper lens acts as the object plane for the lower lens, assuming we want the lower lens to further magnify the image produced by the upper lens. In principle, there's nothing to stop us magnifying as much as we wish.

Don't confuse magnification with resolution.

Above a certain magnification, we will see no more information because other factors limit the image detail and therefore the resolution of the microscope. We'll discuss this point in Section 6.6. We'll also see that there are times when we want to look at an image of the focal plane (because this contains the diffraction pattern). To do this, the back focal plane of the upper lens must become the object plane for the subsequent lenses in the imaging system. When discussing focus we need another convention.

- If the lens is too weak and the image forms below the desired image plane, the image will be out of focus and the lens is said to be *underfocused.*
- If the lens is too strong and the image forms above the image plane, then we say the lens is *overfocused.*

It's very easy to confuse these two terms, unless you think in terms of the vertical frame of the microscope as shown in Figure 6.5.

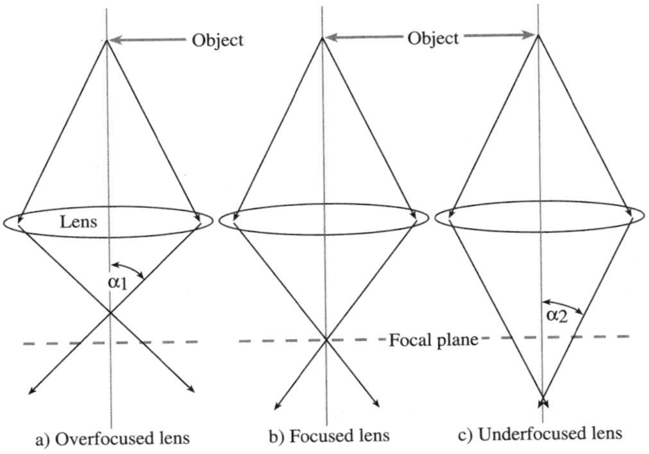

Figure 6.5. (a) Ray diagram illustrating the concepts of overfocus, in which a strong lens focuses the rays before the image plane, and (c) underfocus, where a weaker lens focuses after the image plane. It is clear from (c) that at a given underfocus the convergent rays are more parallel than the equivalent divergent rays at overfocus ($\alpha_2 < \alpha_1$).

6.3. ELECTRON LENSES

Electrons were first successfully focused by Busch in 1927; he used an electromagnet of the sort that Ruska incorporated into the first TEM. Busch also showed that it's possible to focus electrons using electrostatic fields and we've already seen how this works in thermionic electron guns in Chapter 5. In practice, magnetic lenses are superior in many respects, particularly because they are not susceptible to high voltage breakdown. So TEMs use magnetic lenses exclusively and we won't discuss electrostatic lenses further.

6.3.A. Polepieces and Coils

To make a magnetic electron lens you need two parts, and both are drawn schematically in Figure 6.6. First there is a cylindrically symmetrical core of soft magnetic material such as soft iron, with a hole drilled through it. We call this soft iron a "polepiece" and the hole is called the bore of the polepiece. In most lenses there are two polepieces (upper and lower), which can be part of the same piece of soft iron as in Figure 6.6 or may be two separate pieces. The distance between the polepiece faces is called the gap and the bore-to-gap ratio is another important characteristic of such lenses, controlling the focusing action of the lens. Some polepieces are machined to a cone shape and the cone angle is then an important variable in the lens performance.

The second part of the lens is a coil of copper wire which surrounds each polepiece. When we pass a current

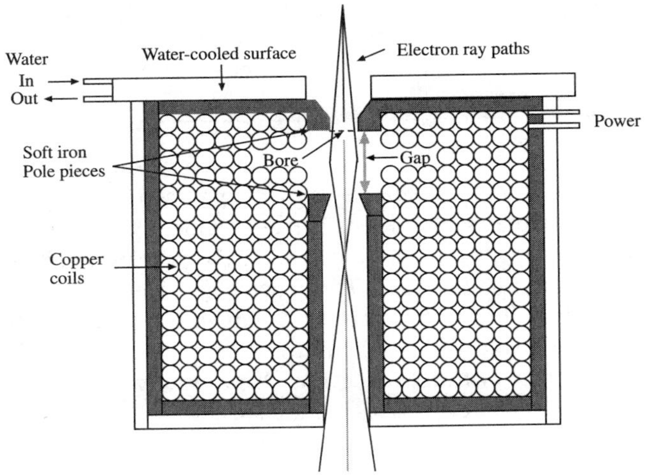

Figure 6.6. Schematic diagram of a magnetic lens. The pole pieces surround the coils and, when viewed in cross section, the bore and the gap between the polepieces are visible. The magnetic field is weakest on axis and increases in strength toward the side of the polepiece, so the electrons are more strongly deflected as they travel off axis.

through the coil, a magnetic field is created in the bore. This field is inhomogeneous along the length of the lens, but axially symmetric. The field lines in a magnetic lens are normal to the ray paths. As you can see, the electron path through the lens is a reasonable approximation to the schematic diagram back in Figure 6.1.

The resistive heating of the coil means that the lenses have to be cooled and a water recirculating system is an essential part of TEM lenses. A real lens removed from the column of a microscope is shown in Figure 6.7.

> Practical hint: *You should be able to get a readout (on your TEM console) of the current through any lens coil and it is a useful thing to know the standard lens currents for your common operating modes such as imaging and diffraction and for various beam sizes.*

6.3.B. Different Kinds of Lenses

The principles that we've just described are incorporated into different kinds of lenses used in the TEM. Most lenses in the microscope are weak lenses with large gaps. Either they act to demagnify the source image onto the specimen, or they magnify the image or the diffraction pattern and project it onto the viewing screen in ways that we'll see in Chapter 9. Typically, these lenses are of the sort shown schematically in Figure 6.6 and an aperture can be introduced into the bore of the lens, as we'll discuss later.

> The objective lens is the most important lens in the TEM, since it forms the images and diffraction patterns that will be magnified by all the other lenses. It is also the most difficult to construct, since the specimen must be located so close to the "plane" of this lens.

The objective lens is a strong lens. Several types exist, depending on the needs of the particular TEM. The most flexible objective lens is that in which the upper and lower polepieces are separated and have their own coil, as shown in Figure 6.8A. This geometry gives the space needed to allow us to insert the specimen and the objective aperture between the polepieces. With this type of polepiece, other instruments such as X-ray spectrometers can have relatively easy access to the specimen. For the same reason, it is straightforward to design specimen holders that do a variety of tasks such as tilting, rotating, heating, cooling, and straining, and this versatility accounts for the popularity of the split polepiece lens in TEMs.

With split polepieces it is possible to make the upper polepiece behave differently than the lower polepiece.

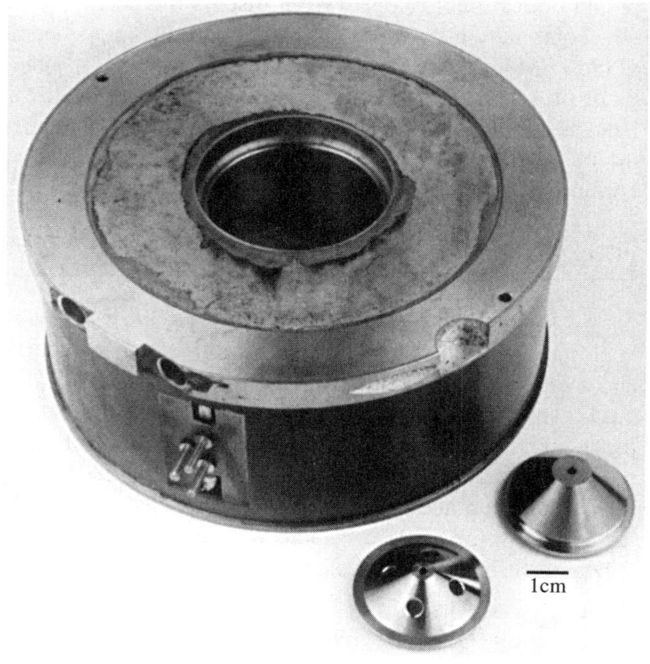

Figure 6.7. A real lens: the cylindrical shape conceals the copper wire coils. The two conical polepieces beside the lens sit inside the central hole in the lens. The three-pin electrical connections provide current to the coil to magnetize the polepieces, and cooling water is circulated in and out of the two holes in the top plate of the lens to dissipate the resistive heat generated.

The most common application of this is to excite the upper objective polepiece very strongly. This kind of lens is ideal for an AEM/STEM because it can produce both the necessary broad beam of electrons for TEM and a fine beam of electrons for AEM and STEM. We'll see how this is accomplished in more detail in Chapter 9.

If high resolution is a major requirement, then we'll see that it is essential to keep the focal length of the objective lens short and this means a very strong lens is needed. This is traditionally accomplished by an immersion lens in which the specimen is dropped into (i.e., immersed in) the center of the lens field as shown in Figure 6.8B. In such a top-entry stage the specimen is surrounded by the objective lens, and so it is a more difficult engineering feat to manipulate, heat, or cool the specimen and it is not possible to get X-ray detectors near the specimen, so analytical microscopy is very inefficient. If the focal length is kept really short to give the highest resolution, then it becomes difficult to tilt the specimen more than a few degrees. So in the highest-resolution TEMs you can't do much apart from imaging and diffraction over a restricted range of tilt (see Chapter 8 on stages). This limitation can be overcome by more recent lens designs such as the snorkel lens, as shown

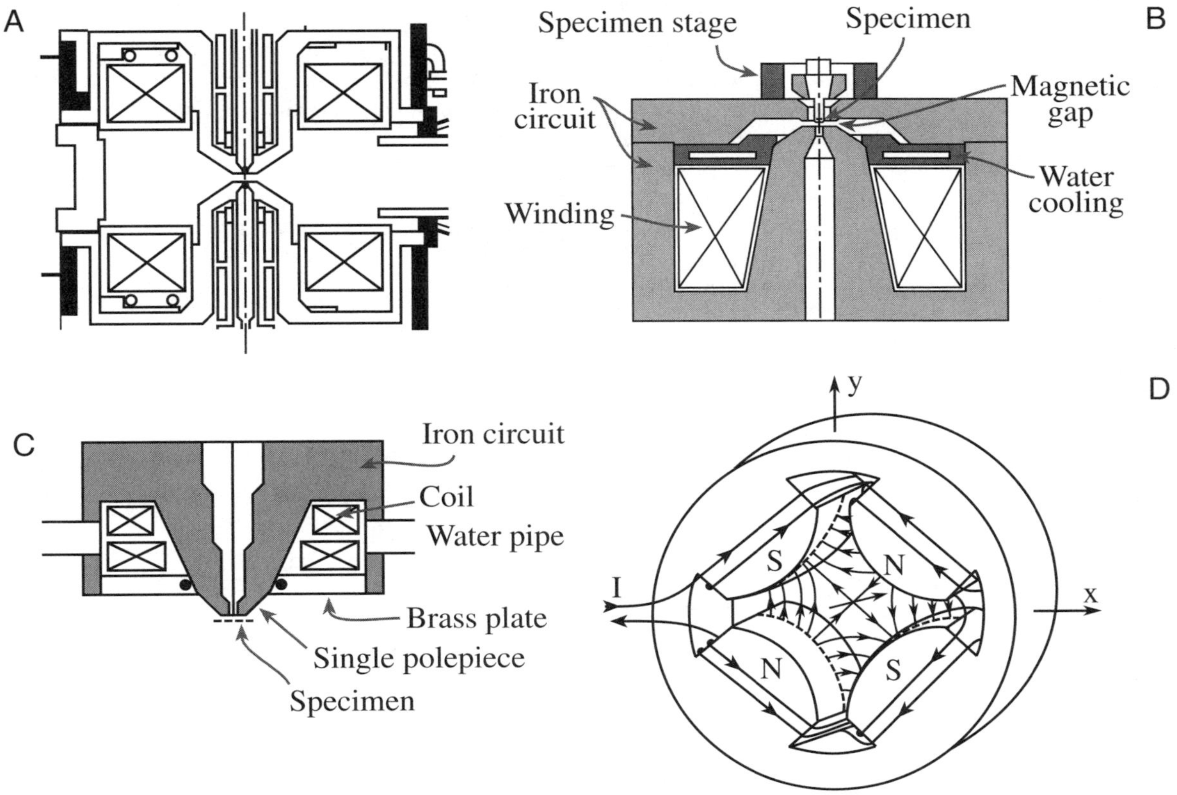

Figure 6.8. A selection of different lenses: (A) a split polepiece objective lens, (B) a top-entry immersion lens, (C) a snorkel lens, and (D) a quadrupole lens.

in Figure 6.8C, which is a single polepiece lens with a small bore to give a strong lens.

The limitations of ferromagnetic polepieces can be overcome using superconducting lenses. We cannot make soft iron polepieces stronger than their saturation magnetization and this limits the focal length and the probe-forming capability of the lens. Superconducting lenses can overcome these limitations, but since a superconductor generates a fixed field it cannot be varied in the same way as a conventional ferromagnetic lens and so it is not very flexible. Nevertheless, there is a lot of interest in these lenses because they are small, they don't need water cooling, and they cool the area around the specimen which both improves the vacuum and helps minimize contamination. They can generate intense fields (>100 T) which are very promising for forming fine probes with high-energy electrons (useful in AEM). Superconducting lenses are so strong that their aberrations (which we'll get to in Section 6.5) are reduced to the level where resolutions < 0.1 nm are feasible and they may be used in the future to construct compact TEM columns.

In addition to these variations on the theme of a single or double polepiece, it is also possible to design a quadrupole and octupole lens in which the focusing action is achieved by four or eight polepieces. Adjacent polepieces are of opposite polarity as shown in Figure 6.8D. These lenses are not used in TEMs as magnifying lenses but they are used to correct lens defects such as astigmatism, and they are used as lenses in electron spectrometers (see Chapter 37). These lenses require less power, and they don't introduce any rotation into the image which, as we'll now show, is a characteristic of standard electromagnetic lenses.

6.3.C. Electron Ray Paths through Magnetic Fields

We need a bit of mathematics to explain how the magnetic lenses actually work. When an electron with charge q ($= -e$) enters a magnetic field of strength $\mathbf{B}$ (Tesla) and an electric field of strength $\mathbf{E}$, it experiences a force $\mathbf{F}$, known as the Lorentz force, which depends on the velocity of the electron, $\mathbf{v}$. All these factors are related through the equation

$$\mathbf{F} = q(\mathbf{E} + \mathbf{v} \times \mathbf{B}) = -e(\mathbf{E} + \mathbf{v} \times \mathbf{B}) \qquad [6.3]$$

where the term in parentheses is a vector cross product. Since we are not applying an electric field within the lens, the resulting (Lorentz) force **F** is a vector normal to **v** and **B**, which are inclined to one another at an angle θ. You can easily work out the relative directions of **E**, **v**, **B**, and **F** using the right-hand rule in which your thumb represents the direction of the force acting on a *positive* charge moving in the direction of the middle finger through a field in the direction of the index finger. So the force on the electron acts in the opposite direction to the thumb.

> Field: Forefinger; Velocity (Speed): Second finger; Thrust: Thumb

Now θ is very nearly 90° for the case of an on-axis electron entering the lens, so we can write

$$F = evB \sin\theta = evB = \frac{mv^2}{r} \qquad [6.4]$$

where r is the radial distance of the electron from the optic axis and m is the mass of the electron.

We can rearrange equation 6.4 to give an expression for

$$r = \frac{mv}{eB} \qquad [6.5]$$

Since v is a relativistic velocity, we should write this equation as

$$r = \frac{\left[2m_0E\left(1 + \frac{E}{2E_0}\right)\right]^{1/2}}{eB} \qquad [6.6A]$$

where m_0 and E_0 are the rest mass and energy of the electron, respectively. This form of the equation allows us to substitute known constants to estimate r (in meters)

$$r = \frac{3.37 \times 10^{-6}\left[V(1 + 0.9788 \times 10^{-6}\,V)\right]^{1/2}}{B} \qquad [6.6B]$$

> For example: if V = 100 kV and the magnetic field B is 1 Tesla, the radius is about 1 mm.

In deriving equation 6.4, we made a rather gross oversimplification. If θ does equal 90°, the electron is traveling straight down the optic axis and is not focused! It is the deviation from θ = 90° that gives the lens effect. The next step therefore is to separate the electron velocity v in a magnetic field into two components, v_1 perpendicular to, and v_2 parallel to, the magnetic field direction **B**, as shown in Figure 6.9, where $v_1 = v \sin\theta$ and $v_2 = v \cos\theta$. The parallel component, v_2, results in motion parallel to the optic axis in the z direction, with $z = v_2 t$, while the perpendicular component produces circular motion with radius given by equation 6.5. Hence we see the very important result:

> The electron spirals through the lens field with a helical trajectory.

The period of rotation (T_c) through the field gives rise to a so-called "cyclotron frequency" ω

$$\omega = \frac{2\pi}{T_c} = \frac{eB}{m} \qquad [6.7]$$

From these various relationships, we can calculate the complete ray paths through the lens. The most important equations are called the "paraxial" (near-axis) ray equations, which describe both r and the angle of rotation (θ) about the axis as the electron moves down the axis in the direction z, rotating under the influence of the cylindrically symmetrical field, B. These equations, which neglect electron trajectories far off axis, are derived in texts on electron optics. The account by Hawkes (1972) is particularly clear if you have an interest in the physics of electron lenses. As Hawkes succinctly states, "a straightforward, but quite lengthy calculation yields"

$$\frac{d^2r}{dz^2} + \frac{\eta^2 B^2 r}{2V^{1/2}} = 0 \qquad [6.8]$$

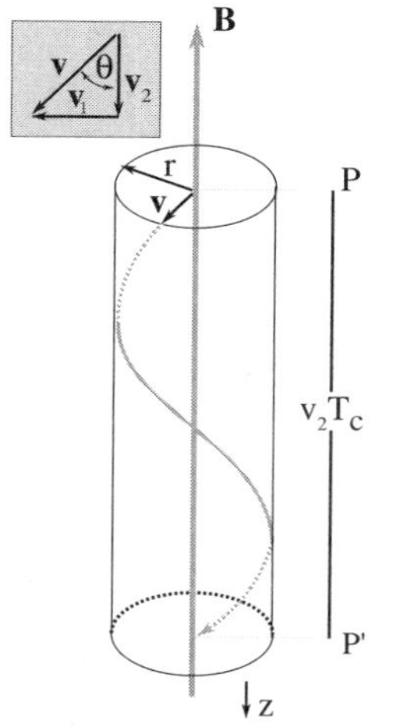

Figure 6.9. Electron trajectories in a homogeneous magnetic field, strength **B**. The electrons have velocity components parallel and perpendicular to the field, so long as they are not traveling at 90° to the direction of **B**. The Lorentz force causes electrons passing through point P on the optic axis to spiral through the field and intersect the axis again at P'. The electron's helical path defines the cyclotron radius, r.

$$\frac{d\theta}{dz} = \frac{\eta B}{2V^{1/2}} \quad [6.9]$$

where V is the accelerating voltage of the microscope and η is $(e/2m_0c^2)^{1/2}$. You can see from equation 6.8 that the rate of change of r is smaller for more energetic electrons and larger for more intense field strengths. From equation 6.9 the rate of angular rotation increases with increasing field strength and decreases for more energetic electrons.

> When the energy of the electrons increases, we must use stronger lenses (larger **B**).

> When we increase **B**, the pitch of the helical path becomes steeper.

Both of these conclusions are intuitively obvious, but the implication is often missed. When we change our operating kV, we change the lenses in the microscope! (Think what this would mean in a visible-light microscope.) Thus the calibration of the microscope and lens "constants" change as we change the accelerating voltage.

While all these ray equations are approximations, they form the basis of more detailed mathematical models of electron motion through lenses. The more complete models are used in a computer program (Munro 1974), which simulates the effects of new lens shapes, bore/gap ratios, etc., and has permitted significant advances in the design of lenses to meet the more stringent demands of modern TEMs. We also use this concept ($\theta \neq 90°$) to explain the effect of spherical lens aberrations on resolution in Section 6.5.A.

6.3.D. Image Rotation and the Eucentric Plane

The electrons follow a helical path as they traverse the field along the axis of the lens. This rotation is rarely shown on standard ray diagrams. Its effects are seen in the routine operation of the TEM because the image, or diffraction pattern, rotates on the viewing screen as you try and focus or if you change magnification. This rotation may require calibration; as we'll see later, the manufacturer may have compensated for it by including an extra lens.

We've already seen in Figure 6.4 that if we change the strength of the lens, the position of the focal plane and the image plane will also change. Because of this, we have to define a standard object plane for the main imaging lens of the microscope and we call this the *eucentric* plane. Your specimen height should always be adjusted to sit in the eucentric plane because an image of an object in this plane will not move as you tilt the specimen. All other planes in the imaging system are defined with reference to the eucentric plane. If your specimen is in the eucentric plane, then the objective lens strength is always the same when the image on the screen is in focus. We'll return to this point later in the book.

6.3.E. Deflecting the Beam

There are many occasions during the operation of the TEM when we want to deflect the beam entering the lens. We may wish to deflect the beam laterally off axis or tilt it to a certain angle with respect to the optic axis. In STEM, these operations are essential to the whole process of forming a scanning image. It is also exceedingly useful in AEM to be able to blank the beam, i.e., deflect it off axis so it goes into a Faraday cup to measure the current, or to prevent the beam from hitting the specimen when no useful spectroscopic data are being gathered. The way we do this is to apply an electromagnetic field to tilt or traverse the beam, or an electrostatic field to blank it. Electromagnetic scan times are of the order of milliseconds while electrostatic blanking can occur in fractions of a microsecond. Although we are drawing the lens as having zero thickness along the optic axis, the magnetic field actually acts over a length L. The angle of deflection ε is (for small ε)

$$\varepsilon = \frac{eLB}{mv} \quad [6.10]$$

From this equation we can show that to tilt the beam by 5° we need a coil carrying about 0.2 A and ~100 turns applied along a length of 1 cm, giving a field of 0.01 T. For electrostatic blanking we need about 20 kV/cm.

6.4. APERTURES AND DIAPHRAGMS

We mentioned earlier that an aperture is often inserted into a lens. The aperture limits the collection angle of the lens as shown schematically in the diagram in Figure 6.10, and such an aperture in the objective lens thus allows us to control the resolution of the image formed by the lens, the depth of field and the depth of focus, the image contrast, the collection angle of the electron energy-loss spectrometer, the angular resolution of the diffraction pattern, etc. Physically, the aperture may reside above, in, or below the plane of the lens as we draw it in ray diagrams. Apertures can also perform other functions which we'll come across later, such as protecting the specimen from stray radiation

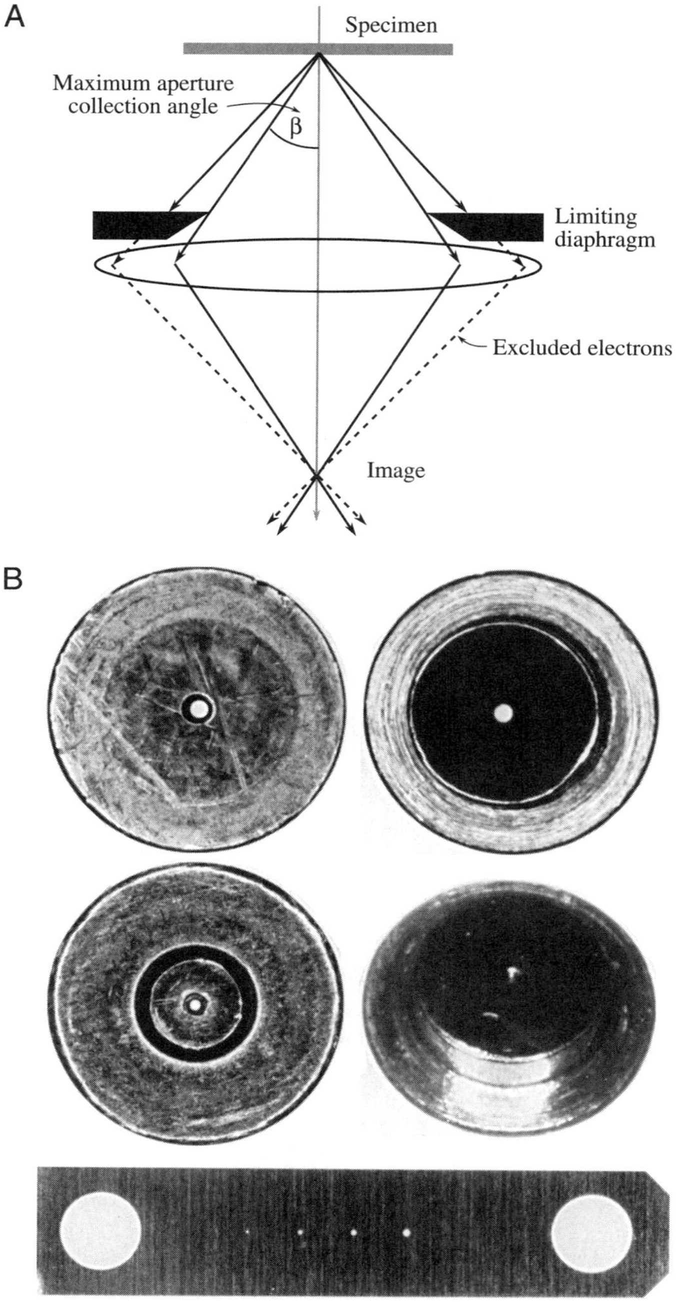

Figure 6.10. (A) Ray diagram illustrating how a diaphragm restricts the angular spread of electrons entering the lens. Only electron paths less than a semiangle β subtended by the aperture at the object are allowed through the lens (full ray paths). Electrons from the object scattered at angles >β are stopped by the diaphragm (dashed ray paths). (B) A selection of diaphragms: the top and middle left are upper and lower views, respectively, of a conventional objective diaphragm; the top/middle right are views of a "top-hat" (thick) C2 diaphragm; below is a metal strip containing several apertures. Each diaphragm is ~3 mm across.

in the illumination system. So they are really important. Usually the apertures are circular holes in metal disks and the disks are made of either Pt or Mo, which are both refractory metals.

> *A quick word on terminology:* While the aperture is the hole in the disk, the metal surrounding the aperture is called the diaphragm (like the variable iris diaphragm in an optical microscope or your camera). We use the aperture to allow certain electrons to pass through the lens and exclude others by causing them to hit the surrounding diaphragm. This "aperture/diaphragm" wording, while strictly correct English, is a bit cumbersome, and microscopists tend to be lazy and just use the term "aperture" in both the correct sense of a hole, but also incorrectly to describe the action of the diaphragm. So we might say that "the objective aperture was used to exclude high angle scattered electrons from the image" or, as we said above, "the aperture protects the specimen from stray radiation" while, strictly speaking, the diaphragm did the excluding and protecting. We'll try to be both consistent and correct in our usage of both terms, but sometimes the precise terminology is awkward.

Diaphragms come in several forms, depending on their function and the particular microscope. They can be either individual disks with a varying diameter aperture, or they can be a series of different apertures in a single metal strip (as shown in Figure 6.10). The diameter can be as small as 10 μm, which is about the smallest circular aperture we can make consistently, up to ~ 0.3 mm (300 μm). Usually the individual diaphragms or the strips are about 25–50 μm thick, but if their job is also to prevent X-rays from hitting the specimen they may be several millimeters thick, which means they can be quite expensive if they're Pt.

Often the diaphragm collects contamination caused by the electron beam cracking residual hydrocarbons in the vacuum (as we describe in Chapter 8). The contamination tends to gather on the edges of the aperture, destroying its circular shape, and this causes astigmatism. So the diaphragms need occasional cleaning, which can be done by heating them to red heat in the central blue part of a butane flame. In some TEMs, this problem is eliminated by making the diaphragm of very thin metal foil (e.g., Au or Mo) because the foil gets hot in the electron beam and boils off any contamination. But such self-cleaning diaphragms are delicate and often crack, thus allowing electrons through other gaps, which defeats the object of the exercise of producing a well-defined aperture.

A safety note: *X-rays with energies up to the beam energy are generated within the lens wherever the electron beam hits a surface (particularly the limiting diaphragm). So substantial and carefully designed lead shielding is incorporated into the column of the TEM to prevent irradiation of the operator. Obviously, it could be very dangerous to tamper with the lenses or diaphragms of the microscope in any way and only qualified service engineers should be allowed to dismantle or in any way take apart and repair the lenses.*

6.5. REAL LENSES AND THEIR PROBLEMS

It might appear from what we've discussed so far that the analogy between electromagnetic lenses and convex glass lenses is complete, but that is not so. Over the 300 years since van Leeuwenhoek first constructed a light microscope, glass lenses have developed to a point where perfect lenses can be fabricated. In the 70 years since Busch's first magnetic lens, we haven't progressed so far and our lenses are very imperfect. We've already compared the best electromagnetic lens to the equivalent of using the bottom of a Coke bottle as a magnifying glass. Another common description is that if your own eye lens was as good as our best electromagnetic lens, then you'd be legally blind! So we have to modify all the ideal ray diagrams we've drawn to take into account the imperfections of the lenses. These imperfections all limit the resolution of the microscope but, paradoxically, help us to get better depth of focus and depth of field from the microscope.

There are ten kinds of lens defects (see Reimer 1993) and at one time or another all their effects can be seen. In practice, however, we don't need to know about all of them and we'll emphasize the ones that limit the microscope performance in substantial ways. These comprise spherical aberration, chromatic aberration, and astigmatism.

6.5.A. Spherical Aberration

The term "spherical aberration" has almost entered the popular vocabulary since its presence was discerned in the main optics of the Hubble Space Telescope (unfortunately after launch). This defect is caused by the lens field acting inhomogeneously on the off-axis rays. The further off axis the electron is, the more strongly it is bent back toward the axis. As a result, a point object is imaged as a disk of finite size, which limits our ability to magnify detail because the detail is degraded by the imaging process. Ultimately, it is this defect that limits the resolution of modern TEMs so we need to examine it carefully.

The effects of spherical aberration are shown in Figure 6.11. A point object P is imaged as an intense central bright region with a surrounding halo in the Gaussian image plane P′. The smallest image of the point source is obtained in a plane just above the Gaussian image plane. This image goes by the delightful term of the "disk of least confusion" or sometimes the "disk of minimum confusion." Spherical aberration is most important in the objective lens because it degrades the image that we view in a TEM. Also, it is equally important in the condenser lenses in an AEM or STEM which we use to form the finest beam with the most current. What we can accomplish is almost always limited by spherical aberration.

From Figure 6.11 you can see why we use the term "spherical" to describe the aberration. The effect is to take

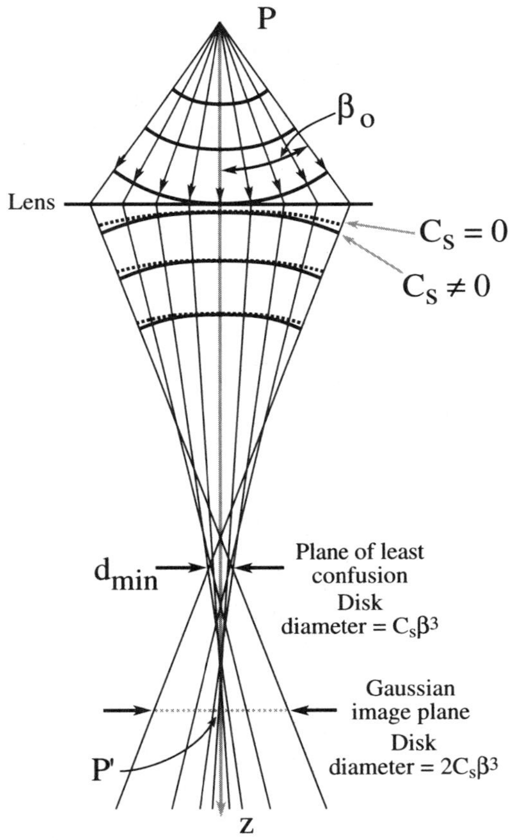

Figure 6.11. Spherical aberration in the lens causes wavefronts from a point object P to be spherically distorted. The point is thus imaged as a disk with a minimum radius in the plane of least confusion and a larger disc at P' in the Gaussian image plane.

the curved wavefronts and further curve them. Now if you go back and look at Figure 6.9, you'll see that electrons traveling through a point P on axis intersect the axis again at point P', where the distance PP' is given by (Reimer 1993)

$$\begin{aligned} PP' &= v_2 T_c = v T_c \cos\theta = 2\pi \frac{mv}{eB}\left(1 - \frac{\theta^2}{2} + \cdots\right) \\ &= L_0\left(1 - \frac{\theta^2}{2} + \cdots\right) \end{aligned} \qquad [6.11]$$

In this relationship $L_0 = PP_0'$, where P_0' is the Gaussian image of the point P for very small θ (i.e., paraxial conditions). As θ increases, the distance PP' decreases because of spherical aberration and we can write

$$PP' = PP'_0 - \Delta z \qquad [6.12]$$

where $\Delta z = 0.5L_0\theta^2$. Thus we get an expression describing the error, δ, in the Gaussian image position due to spherical aberration

$$\delta = \Delta z \tan\theta \sim \Delta z\theta = 0.5L_0\theta^3 \qquad [6.13]$$

So the diameter of the Gaussian image of a point *formed by paraxial rays* is given by this expression, which we usually write as

$$\delta = C_s\theta^3 \qquad [6.14]$$

where C_s is a constant for a particular lens called the spherical aberration coefficient.

Now this equation is for paraxial rays only, and in a real microscope the apertures are usually large enough so that paraxial conditions don't apply. As a result the spherical aberration error in the Gaussian image is broadened to a diameter of $2C_s\theta^3$. We'll see in a while that this equation is *very* important because of its effect on the resolution of the TEM, and sometimes it is written $2C_s\theta^3 M$ when referred to the image plane. In a real lens the value of θ in equation 6.14 is replaced by the maximum semiangle of collection of the objective lens aperture, β. We'll use β to define the objective lens semiangle of collection in the forthcoming discussion of resolution. This contrasts with our use of α when discussing beam size in Chapter 5, since α defined a semiangle of beam convergence. Most other TEM texts use α indiscriminately for collection and convergence angles. When we discuss resolution in the TEM later in this chapter, we will use the radius rather than the diameter, and since all discussion of resolution refers back to the object, the magnification term is customarily ignored.

So we end up with an expression for the radius of the spherical aberration disk r_{sph} in the Gaussian plane under nonparaxial (i.e., realistic) conditions, given by

$$r_{sph} = C_s\beta^3 \qquad [6.15]$$

From this derivation you can see that C_s has the dimensions of length. Typically, it is approximately equal to the focal length, which in most TEMs is about 3 mm, but in high-resolution instruments may be well below 1 mm. So one way to minimize this aberration is to put your money into buying a TEM with a short focal length lens.

If you look at Figure 6.11, you will see that the smallest dimension of the cone of rays formed by the lens does not occur at the Gaussian image plane. The smallest dimension is called the disk of least confusion and has a radius of $0.5C_s\beta^3$. As we'll discuss in Section 6.6.C, some texts use this smaller dimension to define the resolution limit imposed by spherical aberration.

6.5.B. Chromatic Aberration

This term is related to the "color" (i.e., wavelength and hence energy) of the electrons. We've assumed the electrons are monochromatic, but they aren't. However, we can now make very good high-tension supplies and the variation of the electron energy is usually smaller than one part in 10^6, which is 0.1 eV for a 100-keV beam. In fact, this is so good that we don't have to worry about chromatic aberration in the illumination system. In TEM images, chromatic aberration could also be safely ignored if we didn't put a specimen into the beam. Unfortunately, this rather essential action creates electrons of a whole range of energies emerging from the thin foil (for reasons we described in Chapter 4). The objective lens bends electrons of lower energy more strongly and thus electrons from a point in the object once again form a disk image (Figure 6.12). The radius r_{chr} of this disk is given by

$$r_{chr} = C_c\frac{\Delta E}{E_0}\beta \qquad [6.16]$$

where C_c is the chromatic aberration coefficient of the lens (like C_s it is a length, approximately equal to the focal length), ΔE is the energy loss of the electrons, E_0 is the initial beam energy, and β is the semiangle of collection of the lens. While ΔE in the incident electron beam is < 1 eV, it is typically 15–25 eV for a good fraction of the electrons coming through a typical thin foil 50–100 nm thick. Chromatic aberration gets worse for thicker foils and better for thinner ones. So you can do something about this aberration—make good thin specimens!

6.5.C. Astigmatism

Astigmatism occurs when the electrons sense a nonuniform magnetic field as they spiral round the optic axis.

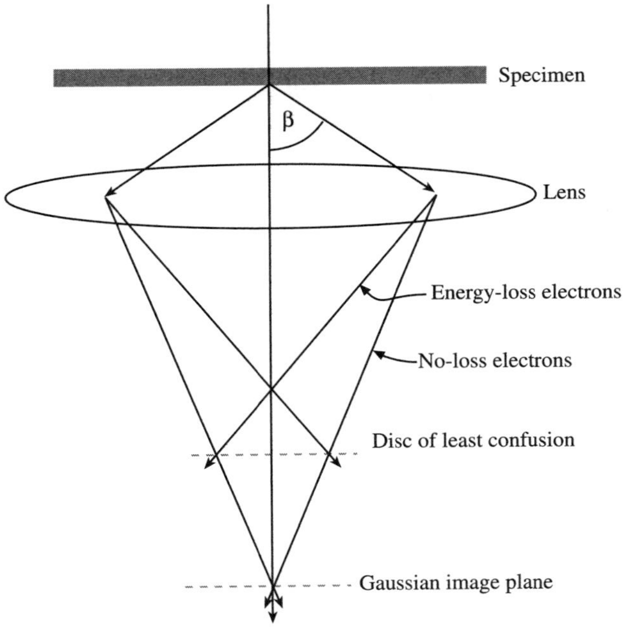

Figure 6.12. Chromatic aberration results in electrons with a range of energies being focused in different planes. Electrons emerging from the specimen with no loss of energy are less strongly focused than those that suffered energy loss in the specimen, so a point is imaged as a disk.

This defect arises because we can't machine the soft iron polepieces to be perfectly cylindrically symmetrical down the bore. The soft iron may also have microstructural inhomogeneities which cause local variations in the magnetic field strength. Even if these difficulties were overcome, the apertures we introduce into the lens may disturb the field if they are not precisely centered around the axis. Furthermore, if the apertures are not clean, the contamination charges up and deflects the beam. So there is a variety of contributions to astigmatism, which distorts the image by an amount r_{ast} where

$$r_{ast} = \beta \, \Delta f \qquad [6.17]$$

and Δf is the maximum difference in focus induced by the astigmatism. Fortunately, astigmatism is easily corrected using stigmators, which are small octupoles that introduce a compensating field to balance the inhomogeneities causing the astigmatism. There are stigmators in both the illumination (condenser lenses) system and the imaging system (objective lens) and we'll describe how to use them in Chapter 9.

These then are the three major defects in electromagnetic lenses. There are several minor defects, such as barrel and pincushion distortion, which are self-explanatory in terms of how they distort the image. They are occasionally seen at very low magnification where electrons traveling close to the bore of the polepiece appear in the image. Other effects, such as coma, field curvature, etc., we will ignore.

6.6. THE RESOLUTION OF THE ELECTRON LENS (AND ULTIMATELY OF THE TEM)

Another note on terminology: We electron microscopists tend to be rather imprecise in our definition and use of the words "resolution" and "resolving power" and related expressions. We've borrowed these terms from classical light optical microscopy, which is concerned with the imaging of incoherent waves through amplitude contrast. High-resolution performance in the electron microscope is a different matter and involves phase-contrast imaging of reasonably coherent electron waves, so perhaps we shouldn't be surprised if a different usage has developed. But we should at least define the terms we use. Now in light microscopy (Bradbury *et al.* 1989) the term "resolution" strictly applies to the act of displaying fine detail in an *image.* The resolving power of the microscope is the ability to make points which are closely adjacent in the object, distinguishable in the *image.* Now the minimum distance apart of these points in the *object* is the *minimum resolvable distance.* Since electron microscopists customarily talk about the resolution of the microscope in terms of distance in the *object* (usually a few Å), we should then use the term "minimum resolvable distance," but instead everyone says "resolution."

> We will use the term "resolution," but we define it to mean the "minimum resolvable distance" in the object.

Because the lens defects that we've discussed cause a point object to degrade into a Gaussian image with a finite radius (some combination of r_{sph}, r_{chr}, r_{ast}) they limit the resolution of the electron lens, and hence that of the microscope. The image resolution in the TEM is governed by the ability of the objective lens to image the object, while in the STEM the image resolution is governed by how much beam current we can put into a small image of the electron source demagnified onto the specimen. In either case, the aberrations limit the resolution.

6.6.A. Theoretical Resolution

If there are *no* aberrations at all, the resolution of *any* lens (glass, electromagnetic) is customarily defined in terms of the Rayleigh criterion, which we introduced back in Chapter 1. Rayleigh devised a criterion for resolution which is arbitrary in the sense that it is not a fundamental physical rule but more a practical definition. This criterion gives us a figure of merit in terms of the eyes' ability to distinguish separate images of two self-luminous incoherent point sources. A single point source will not be imaged as a point, even if no aberrations or astigmatism are present. The finite size of the lens results in diffraction of the rays at the outermost collection angle of the lens, usually defined by a limiting aperture. This diffraction results in a point being imaged as a disk (called the Airy disk) which has a cross-section intensity profile as shown in Figure 6.13a. This should be familiar to anyone who has encountered basic light optics. Rayleigh stated that if the maximum from one source lies over the first minimum from the other source, as shown in Figure 6.13c, then the overall intensity profile exhibits a dip in the middle at about 80% I_{max}. The eye can discern this dip as two overlapping images, thus indicating the presence of two separate objects. Under these circumstances the distance apart of the two incoherent point sources is defined as the theoretical resolution of the lens r_{th} and is given by the *radius* of the Airy disk

$$r_{th} = 0.61\frac{\lambda}{\beta} \qquad [6.18]$$

Beware! Sometimes the diameter is used (hence the term 1.22 λ/β in Chapter 5) because it is the beam diameter which defines image resolution in SEM and STEM. Any standard text on light optics (which we've already referenced) will derive this criterion if you're interested. Now, strictly speaking, we should not use this equation for electron sources because they are not incoherent, and when dealing with true high resolution a different approach is used (see Chapter 28). But for our purposes here, we will be content with this approximation.

Obviously, then we can get higher resolution (smaller r) if we lower λ or increase β. Unfortunately, microscopists often use the expression "higher resolution" when in fact they mean "better resolution" and this can be confusing since the term "higher" is then associated with a lower number. The vacuum is also "higher" if its magnitude is smaller! The improvement in resolution with lower λ accounts for much of the drive toward intermediate and high voltage microscopes since λ decreases with keV, as we saw back in Chapter 1. But the obvious question is why don't we just increase β (i.e., use a bigger lens aperture or remove it altogether). Well, we could do this if we had perfect lenses, but that isn't the case, and the lens aberrations increase as we increase β.

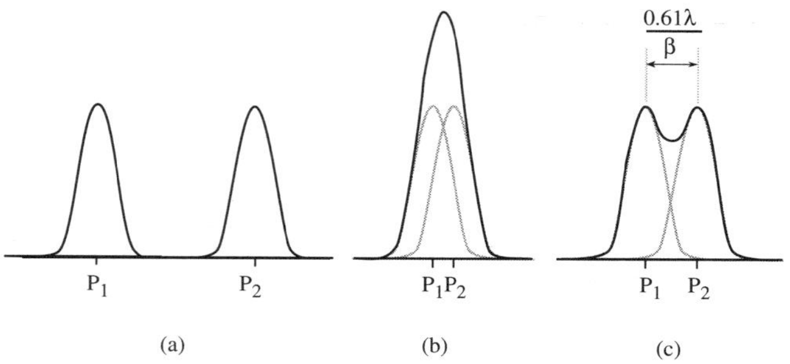

Figure 6.13. (a) The Airy disk intensity profile from two point sources P_1 and P_2 defines the resolution of a lens. In (b) the two Airy disks are so close that they cannot be distinguished, but in (c) the two are separated such that the maximum in the image of P_1 overlaps the minimum in P_2. This is the definition of resolution defined by the Rayleigh criterion.

6.6.B. Spherical Aberration-Limited Resolution (The Practical Resolution)

Let's assume first of all that we have corrected for any astigmatism and our specimen is thin enough that chromatic aberration is negligible. Under these circumstances, the spherical aberration error r_{sph} limits the resolution. Now if you go back and look at equation 6.15 you'll see that r_{sph} increases with the cube of β, a *very* strong dependence. The resolution in the object, then, is given by some combination of the Rayleigh criterion and the aberration error. Hawkes (1972) gives a particularly clear description of how this combination leads to a value for the resolution of the microscope. Since this is very often *the* figure of merit used when investing hundreds of thousands of dollars in a TEM, it is essential that you understand that the definition is not exact. We will follow Hawkes' treatment in detail.

We take the combination of the Rayleigh and spherical aberration disks in quadrature

$$r = \left(r_{th}^2 + r_{sph}^2\right)^{1/2} \qquad [6.19]$$

We can thus find how r varies with β

$$r(\beta) = \left[\left(0.61\frac{\lambda}{\beta}\right)^2 + \left(C_s\beta^3\right)^2\right]^{1/2} \qquad [6.20]$$

Since the two terms vary differently with the aperture collection semiangle β, a compromise value exists when

$$\frac{dr(\beta)}{d\beta} = 0 = -\,-\,2\frac{(0.61\,\lambda)^2}{\beta^3} + 6C_s^2\beta^5 \qquad [6.21]$$

From this equation the optimum (compromise) value of β is given by

$$\beta_{opt} = 0.77 \frac{\lambda^{\frac{1}{4}}}{C_s^{\frac{1}{4}}} \qquad [6.22]$$

(Sometimes, this compromise value is determined approximately by simply equating the equations for r_{th} and r_{sph}. A quick calculation for 100-keV electrons ($\lambda = 0.0037$ nm) for an instrument with $C_s = 3$ mm gives a β_{opt} value of ~15 mrads.)

If this expression for β_{opt} is substituted into equation 6.20, we get a minimum value of $r(\beta)$

$$r_{min} \approx 0.91\left(C_s\lambda^3\right)^{1/4} \qquad [6.23]$$

> This is the expression we want; it gives the practical resolution of the microscope.

Typically, the value for r_{min} is ~0.25–0.3 nm, but the high-resolution instruments have r_{min} ~0.15 nm. So, as we showed back in Figure 1.2, we can resolve rows of atoms, which in most crystalline materials have a separation close to r_{min} (although low index planes in some metals are still below the resolution). It's worth noting that since your eyes can resolve a distance of ~0.2 mm, then the maximum useful magnification of the best high-resolution TEM is ~10^6. Above this magnification, no more detail will be revealed.

Hawkes (1972) reminds us that the decision to add in quadrature back in equation 6.19 was arbitrary, and simply summing r_{th} and r_{sph} is another possible way to determine r_{min} (as we'll see in Section 28.7). But any way you combine the two terms for r leads to expressions of the form of equation 6.22 for β_{opt} and 6.23 for r_{min}. So in many textbooks you will see the use of letters A and B for the constants in these two equations, and often A and B are put equal to unity.

Remember, however, that we ignored any contribution of chromatic aberration. If you have a thick specimen in which most electrons lose 15–25 eV of energy, the chromatic aberration resolution limit given by equation 6.16 will dominate. For example, at 100 keV with β_{opt}, from equation 6.16 the value of r_{chr} is ~2 nm, and you can see all the available information at a magnification of 10^5. Under these circumstances, it doesn't matter what voltage you use or how low your C_s is. If you have thick specimens, then you'll have poor resolution. So how thick is thick? Well, it depends on the voltage and the mean free path for inelastic scatter and many other factors. For good high resolution at 100 keV the specimen should be ~30 nm, while at 300 keV you can probably get away with ~50 nm. A rule of thumb given by Sawyer and Grubb (1987) is that, for biological and polymeric specimens, the resolution limit is about one-tenth the specimen thickness.

6.6.C. Confusion in the Definitions of Resolution

If you're new to the subject, you don't have to read this section because it may confuse you still further, but if you've read other TEM texts you may have noticed discrepancies in the definitions of resolution. Now we used the expression for r_{sph}, measured at the Gaussian image plane. Strictly speaking, it is only under ideal conditions (i.e., if $C_s = 0$) that we should use the Gaussian image as a measure of the resolution limited by the lens. Also, Reimer (1993) further points out that it is only really correct to use the Gaussian image under *paraxial* conditions, that is, with a *very* small objective aperture. As we've already noted, in the TEM β is usually large enough that paraxial conditions do not apply, and so the disk of least confusion is the relevant feature from which to define the resolution, as shown back in Figure 6.11. If this is so, why did we choose the definition of r_{sph} as the radius of the disk in the Gaussian image plane?

The answer to this question is discussed by Hawkes (1972), who notes that defocusing the image slightly, to bring the disk of least confusion to the image plane, will indeed lead to a decrease in the value of B from 0.91 to 0.43. Hawkes also comments that since this latter value is smaller, manufacturers tend to use it to define the resolution of their instrument! However, this whole treatment of resolution assumes incoherent illumination, which is *not* the case in the TEM. Also, the resolution depends on the contrast in the image and how the lens transfers information from the object to the image. As a result, Hawkes concludes that B ~ 1 (from the Gaussian image) is "a more prudent choice" (i.e., closer to reality) than $B = 0.43$ (from the disk of least confusion) even though the disk of least confusion refers strictly to the conditions operating in the TEM.

> *Beware!* Electron microscopists are generally rather careless in their definition of the term (r_{sph}) that they use to describe the effects of spherical aberration on resolution. We use the Gaussian image radius referred back at the object plane, i.e., $r_{sph} = C_s\beta^3$.

Part of this confusion arises out of an uncertainty whether to use the Gaussian image or the disk of least confusion for the definition of r_{sph} and it seems to be a matter of opinion which is more correct. Fortunately, it doesn't really matter too much since, in the end, the choice only alters the value of the constants *A* and *B*, which are often approximated to unity. For example, the value of *A* will depend on exactly which of the several quoted expressions was used for r_{sph}, i.e., if there was 0.25, 0.5, or 1 in front of $C_s\beta^3$. After these various terms are fed into equations and the value of β_{opt} is extracted, *A* only varies by about ±15%. A small variation in *B* will occur also, for the same reason.

We have consistently used the radius of the Airy disk and the radius of the aberration error. Obviously, it doesn't really matter whether you use the radius or the diameter, so long as you are consistent. Occasionally, however, you will find the Airy disk *radius* is used in combination with the *diameter* of the disk of least confusion or the Gaussian image, so this also contributes much to the discrepancy between various TEM texts.

6.7. DEPTH OF FOCUS AND DEPTH OF FIELD

Because of the poor lens quality we have to use small apertures to minimize their aberrations. This generally means that we cut out many of the electrons that would otherwise be gathered by the lens. Fortunately, our electron sources are so intense that we can live with substantially reduced beam currents hitting our specimen. In fact there are advantages to using small apertures, despite the price we pay in image intensity and resolution. These advantages come in the form of better depth of focus and better depth of field. These terms can be confusing and, once again, the TEM literature is variable. So we need to go back to light optics to find the correct definition of these terms.

Basically, we are trying to find out how much of the object (the specimen) is in focus at the same time and over what range the image is in focus. (This latter question is irrevelant in SEM and STEMs where we don't use lenses to form the image, so both terms are equivalent.) In TEM both terms are important.

The depth of field, D_{ob}, is the depth of sharpness in *object space.* It's the distance along the axis on both sides of the object plane within which the object can be moved without detectable loss of sharpness in the image. The depth of focus, D_{im}, is the depth of sharpness in the *image plane,* i.e., the distance along the axis on both sides of the image plane within which the image appears sharp (assuming the object plane and objective lens are fixed).

We can derive expressions for these definitions using Figure 6.14. Imagine that ray 1 originates at the highest point up the column where the object can appear to be in focus within the resolution, and that this ray arrives at the furthest point down the column where the image can appear to be in focus. Ray 2 represents the other extreme but travels at the same inclination to the optic axis. If these two rays appear to come from the same point (to within the resolution of the lens) the distances d_{ob} and d_{im} correspond to the smallest distances which we can resolve in the object or image, respectively. Now we can show that angles α_{im} and β_{ob}, which are both small, are given by

$$\alpha_{im} \sim \tan \alpha_{im} = \frac{\frac{d_{im}}{2}}{\frac{D_{im}}{2}} \qquad [6.24]$$

and

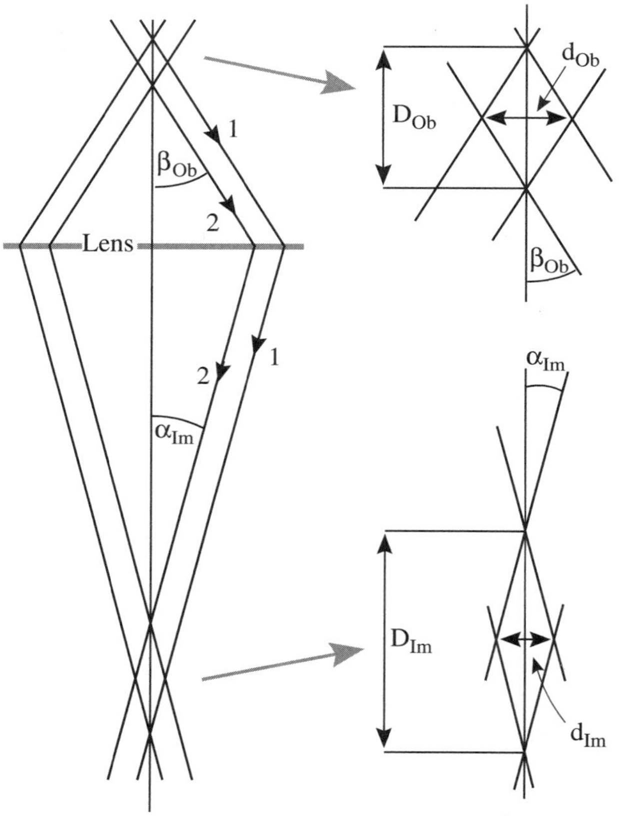

Figure 6.14. The definition of the depth of field and the depth of focus. Rays 1 and 2 represent the extremes of the ray paths that remain in focus when emerging ± $D_{Ob}/2$ either side of the specimen. The same rays define the depth of field over which the image is in focus ± $D_{Im}/2$ either side of the image plane. The resolution in the object is d_{Ob} and that in the image is d_{Im}.

$$\beta_{ob} \sim \tan\beta_{ob} = \frac{\frac{d_{ob}}{2}}{\frac{D_{ob}}{2}} \qquad [6.25]$$

The angular magnification is thus

$$M_A = \frac{\alpha_{im}}{\beta_{ob}} \qquad [6.26]$$

and the transverse magnification (which we simply call the magnification) is

$$M_T = \frac{d_{im}}{d_{ob}} \qquad [6.27]$$

If these two magnifications are related in the usual way by

$$M_T = \frac{1}{M_A} \qquad [6.28]$$

then we can say that the depth of focus is given by

$$D_{im} = \frac{d_{ob}}{\beta_{ob}} M_T^2 \qquad [6.29]$$

and the depth of field is

$$D_{ob} = \frac{d_{ob}}{\beta_{ob}} \qquad [6.30]$$

So we get a much increased depth of focus and field by using small apertures which reduce β. Notice that for a correct calculation of either D_{ob} or D_{im} you must be careful to select the right value of β. Under different circumstances the limiting semiangle is defined by the illumination aperture β (in the C2 lens) or the objective aperture α_o (in the objective lens). In thin specimens, which scatter weakly, most electrons emerge from the specimen in a cone closer to that defined by α_{im}, which is often small (~10^{-4} rad). In a thicker, more strongly scattering specimen, the objective aperture defines the angle and α_o is usually 10^{-2} rad.

For a collection semiangle, β_{ob}, of 10 mrad and a d_{ob} of 2 Å, equation 6.30 tells us that the depth of field will be 20 nm, i.e., a specimen of this thickness can all be in focus at the same time. If you only need 2-nm detail in your image, then you can use a specimen which is 200 nm thick and it will all be in focus.

If we want to see detail at the 2Å level, we need to use a magnification of about 500,000×. Equation 6.29 tells us that, under these conditions, the depth of focus will then be 5 km! If we only need to see 2 nm, we can use a magnification of 50,000× and the depth of focus is 5 m. In either case, we have tremendous latitude in where we put the photographic plate (or other recording media) because it would still record a focused image many meters either side of the screen. This explains why we can use a 35-mm camera which often sits below the final projector lens, and still get a focused image with a TV camera well below the plate camera. In fact, the TEM image would be in focus on the floor under the microscope if you projected it there but M_T would be different!

CHAPTER SUMMARY

We've introduced you to the principles of how an electromagnetic lens works, and how we describe its functions via simple ray diagrams. There are two principal operations: either we use the lens to create an image of an object, or we use it to bring parallel rays to a focus. We'll see in later chapters that the former operation is used to create magnified images of the specimen on the screen of the TEM and also used to create fine electron probes (demagnified images of the electron source) at the plane of the specimen in a STEM or SEM. The latter operation is used to create diffraction patterns in the back focal plane of the objective lens.

Our lenses are rather abysmal in their performance, resulting in the need for small limiting apertures. The lens aberrations limit the resolution of the microscope, and we need an optimum aperture to get the minimum resolution. The small apertures cut down the electron beam intensity, but also give us remarkable depth of focus and depth of field in our images and specimen, respectively.

Points to be wary of when reading about definitions of resolution:
Grundy and Jones (1976), Watt (1985), and Sawyer and Grubb (1987) use the Gaussian image radius referred back at the object plane, as we just did, i.e., $r_{sph} = C_s\beta^3$.
Beware: *Edington (1976) implies, and Thomas (1962), Murr (1970) and Hirsch* et al. *(1977) state, that* $C_s\beta^3$ *is the radius of the disk of least confusion, which it is not, since by definition it must be less than the Gaussian image radius (see Figure 6.11).*
Beware: *von Heimendahl (1980) defines the diameter of the disk of least confusion as* $C_s\beta^3$*, which is also incorrect.*

> Points to be wary of when reading about depth of field and depth of focus
> *Bradbury* et al. *(1989) give a particularly clear discussion of the topic.*
> *Grundy and Jones (1976), Thomas and Goringe (1979), and Sawyer and Grubb (1987) use the conventional definition given here.*
> *Reimer (1993) uses the term "depth of focus" for the "depth of field," a rare inconsistency!*
> *The terms are used interchangeably in SEM because there is no lens between the object and the image.*

REFERENCES

General References

Grivet, P. (1972) *Electron Optics,* Pergamon Press, New York.

Hawkes, P.W. (1982) *Magnetic Electron Lenses,* Springer-Verlag, New York. A collection of review articles.

Hawkes, P.W. and Kasper, E. (1989, 1994) *Principles of Electron Optics,* **1–3**, Academic Press, New York. Comprehensive but advanced. Volume **3** includes imaging in the TEM.

Klemperer, O. and Barnett, M.E. (1971) *Electron Optics,* Cambridge University Press, New York.

Specific References

Bradbury, S., Evennett, P.J., Haselmann, H., and Piller, H. (1989) *Dictionary of Light Microscopy,* Royal Microscopical Society Handbook #15, Oxford University Press, New York.

Busch, H. (1927) *Arch. Elektrotechnik* **18**, 583.

Edington, J.W. (1976) *Practical Electron Microscopy in Materials Science,* Van Nostrand Reinhold, New York.

Grundy, P.J. and Jones, G.A. (1976) *Electron Microscopy in the Study of Materials,* E. Arnold, London.

Hawkes, P.W. (1972) *Electron Optics and Electron Microscopy,* Taylor and Francis Ltd., London.

Hirsch, P.B., Howie, A., Nicholson, R.B., Pashley, D.W., and Whelan, M.J. (1977) *Electron Microscopy of Thin Crystals,* 2nd edition, Krieger, Huntington, NewYork.

Jenkins, F.A. and White, H.E. (1976) *Fundamentals of Optics,* 4th edition, McGraw-Hill, New York.

Munro, E. (1974) Proc. 8th Int. Cong. for Electron Microscopy, **1**, p. 218, The Australian Academy of Sciences, Canberra.

Murr, L.E. (1970) *Electron Optical Applications in Materials Science,* McGraw-Hill, New York.

Reimer, L. (1993) *Transmission Electron Microscopy; Physics of Image Formation and Microanalysis,* 3rd edition, Springer-Verlag, New York.

Sawyer, L.C. and Grubb, D.T. (1987) *Polymer Microscopy,* Chapman and Hall, New York.

Thomas, G. (1962) *Transmission Electron Microscopy of Metals,* John Wiley and Sons, New York.

Thomas, G. and Goringe, M.J. (1979) *Transmission Electron Microscopy of Materials,* Wiley Interscience, New York.

von Heimendahl, M. (1980) *Electron Microscopy of Materials,* Academic Press, New York.

Watt, I.M. (1985) *The Principles and Practice of Electron Microscopy,* Cambridge University Press, New York.

How to "See" Electrons

7

CHAPTER PREVIEW

If we are studying the structure of a material, when all is said and done, all we have to show for our expensive electron microscope, hours of specimen preparation, and careful alignment, etc., is an image, a diffraction pattern, or a spectrum. These images and diffraction patterns, which are just different distributions of electron intensity, have first to be viewed in some manner. After viewing, we have to decide if we want to save the result for future reference, perhaps so we can print it for a technical report or scientific publication. Since, as we noted in the opening chapter, our eyes are not sensitive to electrons, we have to find ways to translate the electron intensity distributions into visible-light distributions. This chapter will explain how we "see" electrons.

We'll break the process down into two parts: first, detection (and display) of the image, and second, recording of the image. Both these areas are undergoing rapid change because of advances in electronics, and so this chapter will undoubtedly contain anachronisms by the time you read it.

How to "See" Electrons

7

7.1. ELECTRON DETECTION AND DISPLAY

Images and diffraction patterns are different two-dimensional electron-intensity distributions which can be produced by scattering by the same object. We detect and display them in different ways depending on whether we are using a TEM or STEM, as we'll explain in Chapter 9. In a conventional TEM the images and diffraction patterns are static, because the incident beam is fixed, and so we can easily project them onto a viewing screen within the microscope column. TEM images, for example, are *analog* images of electron density variations in the image plane of the objective lens. We cannot manipulate the image or its contrast in any way between the electrons leaving the image plane and being projected onto the viewing screen. We will briefly discuss the properties of the viewing screen. The manufacturer controls the initial choice of screen materials so you might think there's not much need to understand this aspect in any depth. You might be surprised by the limitations you don't need to accept or the improvements which could be made.

When we operate our TEM as a STEM, or we use a dedicated STEM, the image is not static; it is built up over time as the small probe is scanned across the area of interest. Under these circumstances, we detect the electron signals by various types of electronic detection. If we are seeking secondary electron (SE) or backscattered electron (BSE) signals, then these detectors sit in the specimen stage area. If we are seeking to image the same forward-scattered electrons that we view on the TEM screen, the detectors are in the viewing chamber of the TEM. After we've detected any one of these signals, it is usually digitized and *digital* scanning images are presented on a fluorescent screen as an analog image. We often refer to this fluorescent screen as the CRT, which is the acronym for "cathode-ray tube" and a relic from the early days of electron physics.

We should point out that the sequential or serial nature of the scanning image makes it ideal for on-line image enhancement, image processing, and image analysis. The signal from any electronic detector can be digitized and electronically manipulated prior to display on the CRT, in a way that is impossible with analog images. We can adjust the digital signal to enhance the contrast or to reduce the noise. Alternatively, we can store the digital information and process it mathematically. The availability of cheap memory and fast computers permits on-line image processing and the rapid extraction of quantitative data from the scanning image; we discuss all this and more in Chapter 30. Because of developments in computer technology, there is great interest in recording analog TEM images via a TV camera in order to digitize them and charge-coupled device (CCD) cameras are already available for on-line viewing and processing, particularly of HRTEM images.

In attempting to compare the properties of detection and recording devices we often use the concept of the "detection quantum efficiency" or DQE. If the detector is linear in its response, then the DQE is defined simply as

$$\mathrm{DQE} = \frac{\left(\dfrac{S_{\mathrm{out}}}{N_{\mathrm{out}}}\right)^2}{\left(\dfrac{S_{\mathrm{in}}}{N_{\mathrm{in}}}\right)^2} \qquad [7.1]$$

where *S*/*N* is the signal-to-noise ratio of the output or input signal. So a perfect detector has a DQE of 1 and all practical detectors have a DQE <1.

Note on terminology: We use several different terms, often imprecisely, to describe how we "see" electrons. Since our eyes can't in fact "see" electrons, we have to resort to the phenomenon of cathodoluminescence (CL) in order to provide an interface between electrons and our eyes. Any electron display system that we look at relies on CL. The CL process converts the energy of the electrons (cathode rays) to produce light (luminescence). As a result, any electron display screen emits light in proportion to the intensity of electrons falling on it.

- *Light emission* caused by ionizing radiation is *scintillation.*
- The process of *fluorescence* implies *rapid emission.*
- *Phosphorescence* implies that the wavelength and the *delay time* are longer than for fluorescence.

All these terms are used in electron microscopy (interchangeably and often inaccurately) because the "fluorescent" screen is coated with a long-delay phosphor (see Chapter 9).

7.2. VIEWING SCREENS

The viewing screen in a TEM is coated with a material such as ZnS, which emits light with a wavelength of 450 nm. The ZnS is usually modified with impurities to give off green light at closer to 550 nm; hence you'll see screens of different shades of green which, being in the middle of the visible spectrum, is most relaxing for the eyes. As long as sufficient light is emitted, the main requirement of the viewing screen is that the ZnS particle (grain) size be small enough so that the eye cannot resolve individual grains. This means that grain sizes <100 μm are acceptable (although you can see the grain size if you look at the screen through the auxiliary focusing binoculars). Typical screen coatings are made with a ZnS grain size of about 50 μm, although they may be as small as 10 μm for the highest-resolution screens.

As we've seen in Chapter 4, the emission intensity of most signals, including CL, decreases with increasing beam voltage. You would thus expect the light intensity to degrade at higher voltages, but this is offset by the increase in gun brightness. In some HVEMs the small focusing screen support is made of a heavy metal, such as Pt, to try and encourage backscatter and increase screen intensity. Of course, this backscattering will broaden the volume where light is generated and blur the image, so we don't gain very much. In fact most TEMs have very similar screens. Other signals are also given off by the viewing screen, such as X-rays, and whenever you look at the screen you are protected from this lethal radiation flux by lead glass, carefully selected to absorb the radiation to levels at or below ambient background. In HVEMs this can amount to several centimeters of glass, and invariably the optical transmission capabilities are degraded as the glass gets thicker, but obviously we have no alternative if we want to view the screen directly.

A few practical hints about your screen: There isn't much you can do about choosing the best material for the viewing screen since the manufacturer selects it for you, but you can extend its life substantially by taking care to minimize overexposure. The greatest source of screen damage is the intense direct beam that comes through thin specimens and constitutes the central spot in diffraction patterns. You can minimize burning of the screen by (a) only going to diffraction mode with the selected area aperture inserted, (b) only going to diffraction mode with the C2 lens underfocused, and (c) if the spot appears exceptionally intense despite these precautions, then insert the beam stop while you're observing the pattern on the screen.

7.3. ELECTRON DETECTORS

We have several alternatives to the fluorescent screen for detecting electrons. These other electron detectors play a major role in STEMs and AEMs (as well as in SEMs). They are actually essential to the STEM image-forming process that we'll describe in Chapter 9. Such detectors are usually one of two kinds: semiconductor (silicon *p-n* junction) detectors or scintillator–photomultiplier systems. We'll examine the pros and cons of each of these two types and end with a brief comment on CCDs.

7.3.A. Semiconductor Detectors

A full understanding of how semiconductor detectors work requires a fair knowledge of solid-state physics. We'll just give a brief outline of the principles as they affect the use of the TEM.

The semiconductor detector, shown schematically in Figure 7.1, is a doped single-crystal sheet of Si (often inaccurately described as a solid-state detector). We make the Si into an electron-sensitive detector by creating a *p-n* junction beneath the Si surface in one of two ways. In one type of detector, we create the junction by doping the Si (e.g., by ion implantation of *n*-type impurity atoms into *p*-type Si or vice versa). This doping disturbs the equilibrium charge carrier concentration and creates a region across the *p-n* junction that is free of majority carriers. We call this region a "depletion region." A conducting metal layer is evaporated onto both surfaces to provide ohmic contacts. The alternative type of detector is called a surface-barrier

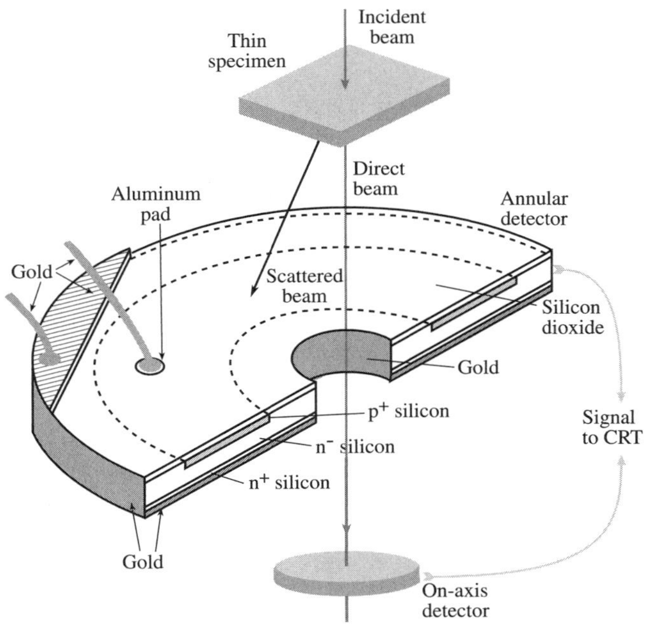

Figure 7.1. Semiconductor detector of the surface-barrier type, shown in a configuration where it would be used to detect high-energy forward-scattered electrons. The direct beam is detected by a small circular detector on the optic axis surrounded by a concentric wide-angle annular detector, which detects any scattered electrons.

detector (or sometimes a Schottky diode) and we fabricate this by evaporating a thin layer of Au on the surface of high-resistivity *n*-type Si, or evaporating Al onto *p*-type Si. This surface layer acts as an electrical contact and also creates a depletion layer and a *p-n* junction just inside the Si.

When we put either of these detectors into a beam of high-energy electrons, most of the beam energy is transferred to valence-band electrons in the Si which are excited into the conduction band; we thus create electron–hole pairs. We can separate the electrons and holes most efficiently by applying an external reverse bias to the detector; that is, we put a negative bias on the *p* side of the junction and a positive bias on the *n* side. In practice, however, so many electron–hole pairs are created at TEM beam energies that an external bias is not usually necessary, and the internal bias of the *p-n* junction acts to separate the electrons and holes. Because the electrons and holes move quite quickly in Si, it takes only a few nanoseconds to gather most of the carriers over an area of about 1 μm^2. So the semiconductor detector is remarkably responsive to the incoming electron signal. The net result of all this is that the incoming electron signal is converted to a current in the external circuit between the surface contacts, as shown in the surface-barrier detector in Figure 7.1.

Since it takes approximately 3.6 eV to produce an electron–hole pair in Si at room temperature, a 100-keV electron can theoretically produce about 28,000 electrons. This represents a maximum detector gain of close to 3×10^4, but in practice there are losses due to electron absorption in the metal contact layer and recombination of the electrons and holes close to the Si surface (in a region called the dead layer), and we actually get a gain of closer to 2×10^4.

These semiconductor detectors are very efficient at picking up and amplifying electron signals. Unfortunately, they have an inherently large capacitance, and so are not very responsive to rapid changes in signal intensity. Such changes are quite likely to occur during the rapid scanning process of STEM imaging. In other words, the detector has a narrow bandwidth (typically 100 kHz), and this is not a good property for a detector which is subject to widely varying signal intensities. We could lower the capacitance by decreasing the detector area, but if we do this, the signal-to-noise ratio will be lowered, and it is this latter factor which ultimately limits the quality of all scanning images.

Semiconductor detectors have several advantages:

- We can easily fabricate them.
- They are cheap to replace.
- They can be cut into any shape, as long as it is flat.

This latter advantage makes them ideal for squeezing into the confines of TEM stages and columns. For example, we can make the semiconductor detector in annular form so that the main electron beam goes through it, but the scattered electrons are very efficiently detected. We thus have a dark-field detector. We can also make detectors that are divided into halves or quadrants. These are very useful for discriminating directional signals such as those coming from magnetic specimens.

There are also some drawbacks to semiconductor detectors:

- They have a large dark current (the current registered when no signal is incident on the detector). This dark current arises from thermal activation of electron–hole pairs, or from light falling on an uncoated detector. Since the detectors in a TEM invariably have a metal ohmic contact, the light problem is minimal. Now we could minimize thermal activation by cooling the detector to liquid-nitrogen temperatures, but that step is impractical and introduces a cold surface into the vacuum which would simply collect contamination, so we live with noise due to the thermal activation.
- Because noise is inherent in the semiconductor detector, its DQE is poor for low-intensity sig-

nals, but rises almost to unity for high-intensity signals.

- The electron beam can damage the detector, particularly in intermediate voltage microscopes. In these circumstances a doped *p-n* detector is less sensitive than the surface-barrier detector, because the depletion region is deeper in the Si.
- They are insensitive to low-energy electrons such as secondary electrons.

Despite these drawbacks, both types of Si detector are far more robust than the alternative scintillator detector, which we will now describe.

7.3.B. Scintillator–Photomultiplier Detectors

A scintillator emits visible light when struck by electrons because of the same CL process that occurs in fluorescent screens. While we are viewing a static TEM image, we want the fluorescent screen to continue emitting light for some time after the electrons hit it, so we use a long-delay scintillator. Of course, when we are using a scintillator to detect rapid changes in signal as in scanning beam imaging, we want the light emission to decay rapidly. So we don't use ZnS in scintillator detectors but rather materials such as Ce-doped yttrium-aluminum garnet (YAG) and various doped plastics and glasses. These materials have decay times on the order of nanoseconds rather than microseconds for ZnS. Once we've converted the incoming electron signal to visible light, the light from the scintillator is amplified by a photomultiplier (PM) system, attached to the scintillator via a light pipe. Figure 7.2 shows a schematic diagram of a scintillator–PM detector set up to detect secondary electrons in a TEM, and the design used in the SEM is essentially identical.

The scintillators that we use in STEMs or SEMs are often coated with a 100-nm-thick layer of Al to reflect any light generated in the microscope and stop it from entering the PM tube, where it would add noise to the signal. If the detector is in the stage of the microscope, this light could come from the specimen itself if it is cathodoluminescent, or it could be light coming down the column from a thermionic source and reflected from the polished surface of the specimen. If you have a scintillator detector in the viewing chamber, then room light may also hit the detector, so you should cover the windows of the viewing chamber.

The advantages of the scintillator–PM system are:

- The gain of the system is very high. The gain for the total detector system is of the order of 10^n, depending on the number (n) of dynodes in the PM. A value of 10^8 is not unusual (compare with 10^4 for the semiconductor detector). This performance is reflected in a typical DQE of close to 0.9 for several commercial scintillators.
- The noise level in a scintillator is low compared with semiconductor detectors, and the bandwidth of the scintillator is in the MHz range. As a result, both low-intensity images and TV-rate images are easily displayed. There is a tremendous practical advantage to TV-rate imaging of digital signals, because such images, when suitably processed and displayed, can be viewed, stored, and recorded under normal room illumination conditions. So you don't have to work in the dark while operating your (S)TEM.

The disadvantages of the scintillator–PM system are:

- The scintillator is not as robust as the semiconductor detector, being more susceptible to radi-

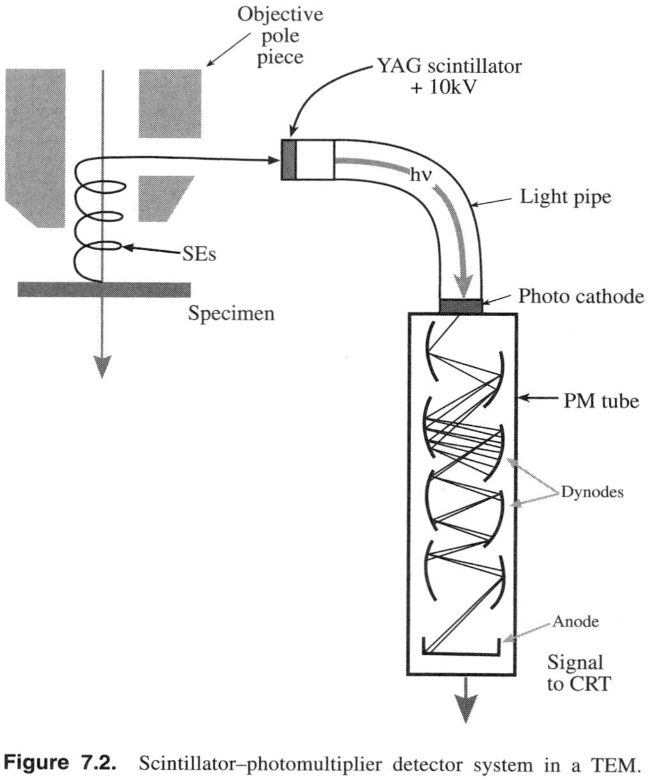

Figure 7.2. Scintillator–photomultiplier detector system in a TEM. SEs from the specimen spiral back up through the polepiece and are accelerated by the high voltage onto the scintillator, generating visible light which travels via fiber optics to a photocathode. There the light is reconverted to electrons. The electron signal is then multiplied by several electrodes in the PM tube.

ation damage, particularly after long-time exposure to the beam.

- The scintillator–PM combination is also substantially more expensive and bulky compared to semiconductor detectors and therefore it does not fit well within the TEM stage, nor is it easily manufactured into multidetector configurations. However, plastic scintillators can be shaped to give large-angle collection, such as in the Robinson BSE detector used in many SEMs.
- The energy conversion efficiency of a scintillator is also rather low (about 2%–20%) compared to a semiconductor detector and typically we only get about 4000 photons per incident 100-keV electron, about 7 times less than the semiconductor detector. This low efficiency is offset by the gain in the PM tube.

On balance, the scintillator–PM detector is preferred over the semiconductor detector for most general electron detection in TEM/STEM systems. However, you must take care to minimize any high-intensity beams that may damage the detector and lower its efficiency. Therefore, you need to take more care when operating scintillator detectors.

7.3.C. TV Cameras and Charge-Coupled Devices

We've already mentioned that you can view the TEM image directly through a TV camera, rather than looking at the fluorescent screen. There are real advantages to doing this for on-line viewing of faint HRTEM images (see Chapter 29), or for recording of dynamic *in situ* events. A standard TV camera is often quite sufficient for this job, although in the US, the TV-image resolution (500 lines/frame) is rather low and a high-resolution camera (1000 lines/frame) is preferred. Video storage and display again requires higher resolution than standard VHS video formats if you want to get the most out of your images. The best TV cameras are CCDs, which are replacing standard plumbicon tube TV cameras.

CCDs are MOS devices that store charge generated by light or electron beams. CCD arrays consist of thousands or millions of pixels which are electrically isolated from each other by creating potential wells under each CCD cell so they can accumulate charge in proportion to the incident beam intensity, as shown in Figure 7.3A. The cells currently can be as small as 6 μm. To create a picture we have to read out the array, which can be done by changing the applied potentials to transfer the charge serially

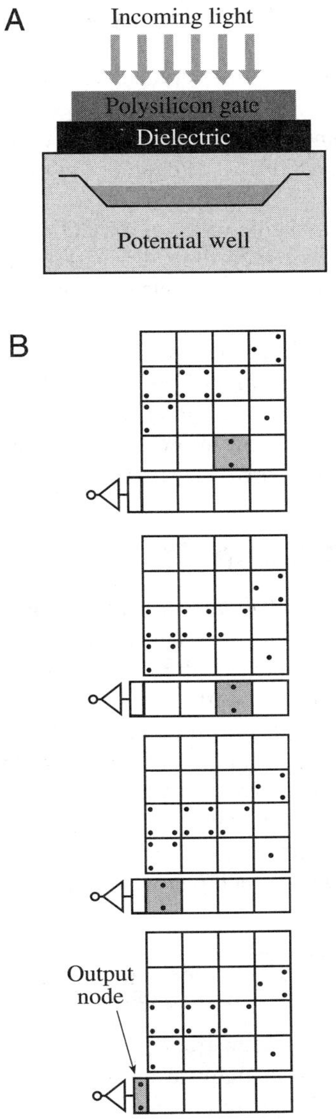

Figure 7.3. (A) A single cell in a CCD array showing the storage of charge in the potential well under one pixel. If we vary the applied potential to rows of pixels in sequence, as in (B), one pixel row is shifted to the parallel register, and is read out pixel by pixel, after which the next row is moved to the parallel register, and so on. The stored charge in each pixel is thus fed into an amplifier and digitized.

from each potential well along a line into an output amplifier as shown in Figure 7.3B. Once all the cells are empty the array can be re-exposed. We thus have a frame time for reading the display which can be as short as 0.01 s, well below standard TV rates (0.033 s). Rather than serial readout, it is also possible to have full frame CCDs in which the whole frame is transferred to an adjacent storage array leaving the main array free to collect a new signal flux. We get an analog output, i.e., a charge current, which we then digitize, usually through an 8-bit A to D converter to give 256 gray levels.

CCD arrays have several advantages:

- When cooled, they have a very low noise and a good DQE (>0.5) even at low input signal levels.
- The dynamic range of a CCD is high, making it ideal for recording diffraction patterns which can span an enormous intensity range.

The major disadvantages are the speed and the expense. These devices can also be used as two-dimensional arrays for parallel-collection electron energy-loss spectrometers rather than the more conventional one-dimensional silicon diode arrays described in Chapter 37.

7.3.D. Faraday Cup

In conventional TEM there isn't much need to know the beam current, but in the AEM it is essential since there is often a need to compare analytical results obtained under identical beam current conditions. A Faraday cup is a detector that simply measures the total electron current in the beam. We don't use it for any imaging process, but rather as a way of characterizing the performance of the electron source, as we saw in Chapter 5. Once the electrons enter the Faraday cup, they cannot leave except by flowing to ground through an attached picoammeter that measures the electron current.

> A Faraday cup is a black hole for electrons.

You can easily construct a Faraday cup to go in an SEM, but it is more difficult to design one that fits in the stage of a TEM. Fortunately, some manufacturers now incorporate a Faraday cup in the specimen holder. You can measure the current by deflecting the beam into the cup or partially extracting the holder (Figure 7.4B). These cups are not ideal because they don't trap all the electrons. A dedicated Faraday-cup holder is shown in Figure 7.4A. The entrance aperture is small and the chamber is relatively deep and lined with a low-Z material to minimize backscatter. If you tilt it slightly, the electrons have little chance of being scattered directly back. With such a holder you can only find the hole if you can image the upper surface with SE or BSE detectors, and if these are not available then you must have a cup with a hole in the lower surface too. When the cup is not tilted, the electrons go straight through; if you tilt the cup, then all the electrons are trapped as shown in Figure 7.4A. The way to ensure that you are measuring the maximum current is to look at the picoammeter reading as you tilt the cup.

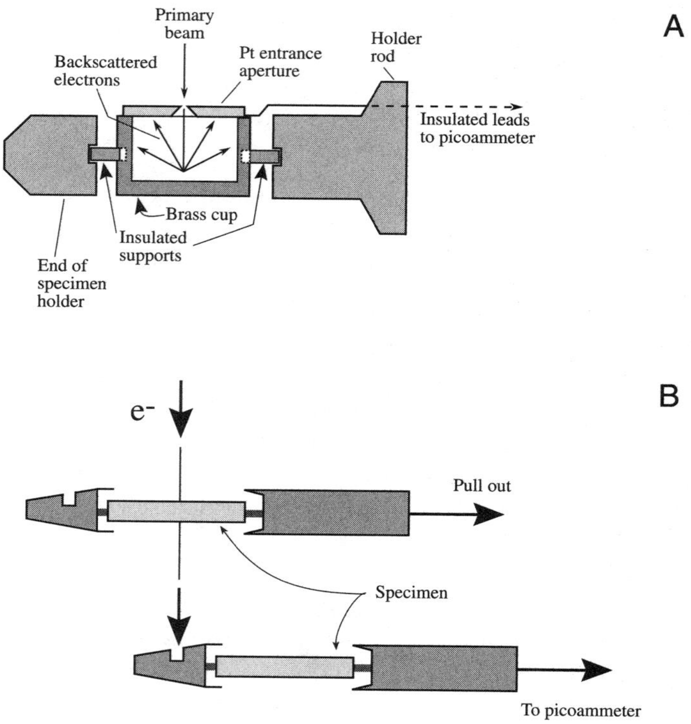

Figure 7.4. (A) Schematic diagram of a Faraday cup in the end of a side-entry specimen holder. The entrance aperture has to be found using SEs or BSEs. In (B) the holder is retracted slightly so the electrons fall into a cup on the tip of the holder. The electron current is measured as it goes to ground through a picoammeter attached to the outside of the holder.

If you don't have a Faraday cup, it is possible to get an approximate reading of the current by just measuring the current through an insulated line from a bulk region of the specimen and correcting for electron backscatter. Backscattering is independent of the accelerating voltage and approximately linear with atomic number up to about $Z = 30$. For example, the backscatter coefficient for Cu is about 0.3 and for Al it is about 0.15. It is also possible to deflect the beam onto the last beam-defining diaphragm and measure the current via an insulated feed-through (also correcting for backscatter).

7.4. WHICH DETECTOR DO WE USE FOR WHICH SIGNAL?

As we mentioned at the start of the chapter, the principal electron signals that we can detect are the forward-scattered electrons, which as we'll see in Chapter 9 are the most common TEM images, and the BSE and SE signals from the beam-entry surface of the specimen.

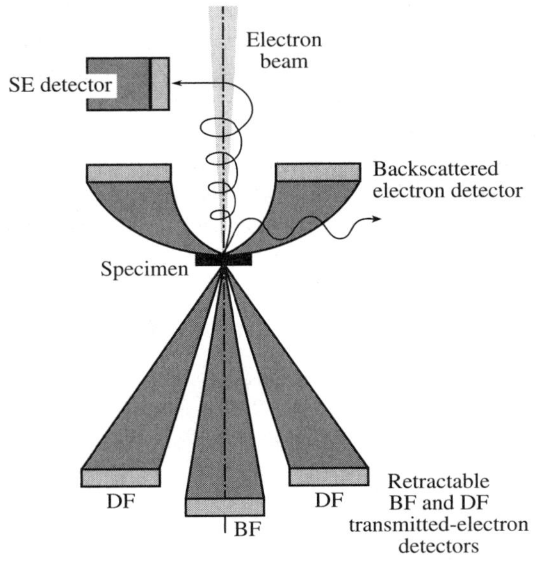

Figure 7.5. The various electron detectors in a STEM. Scintillator–PM detectors are invariably used for SE detection and semiconductor detectors for the BSE. The on-axis and annular forward-scattered detectors may be either type, depending on the microscope.

Semiconductor detectors are only sensitive to electrons with sufficient energy (>5 keV) to penetrate the metal contact layer. So we use these detectors mainly for *high-energy* forward-scattered imaging and *high-energy* BSE imaging. Because of the surface contact layer we don't use semiconductor detectors for *low-energy* SEs and a scintillator–PM system is required. Remember that the scintillator may also be coated with Al to prevent visible light from generating noise. This coating would also prevent low-energy SEs from being detected. So for SE detection, either there must be no coating, or the electrons must be accelerated to an energy high enough to penetrate the coating; we achieve the latter by applying a high kV (>10 kV) positive bias to the scintillator.

The capacitance is relatively high for semiconductor detectors, so they are not the detector of choice in dedicated STEMs where high scan-rate TV images are the normal viewing mode, i.e., where you need a quick response. The scintillator–PM system is again preferred under these circumstances. As most microscopes move toward TV-rate display of scanning images it is likely that the scintillator–PM will be used increasingly for forward-scattered TEM imaging. Semiconductor detectors may only be used for BSEs, which is not a major imaging mode in TEMs. A summary of all the various electron detectors in a TEM/STEM is given in Figure 7.5.

Sometimes we examine specimens which themselves exhibit cathodoluminescence under electron bombardment. We'll discuss why this is done in Chapter 31. A mirror is used to focus the light into a scintillator–PM system, and one such design is shown in Figure 7.6. This setup effectively prevents detection of all other signals, including X-rays, because the mirror occupies all the free space in the TEM stage. So you have to dedicate the TEM to CL detection alone and ignore other signals.

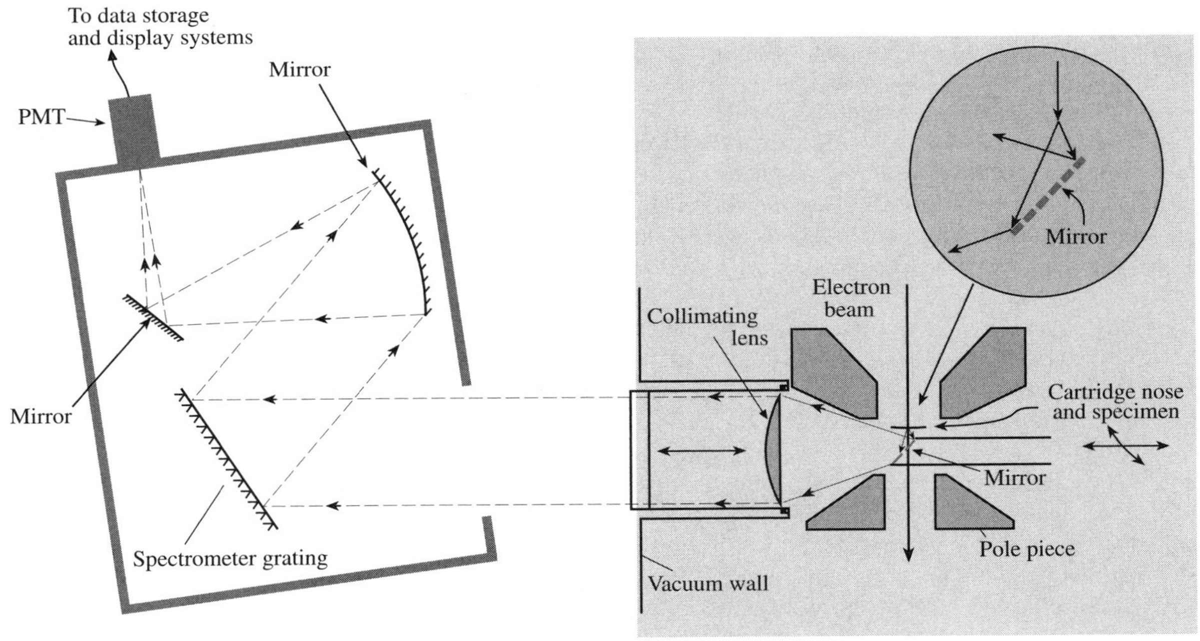

Figure 7.6. Cross section of a mirror detector below a thin cathodoluminescent specimen that collects light and focuses it into a spectrometer–PM system. The CL signal is usually very weak and so the detector has to be as large as possible, and it takes up much of the free volume in the TEM stage.

7.5. IMAGE RECORDING

7.5.A. Photographic Emulsions

Although film is the oldest recording medium, it still retains sufficient advantages that we continue to use it in virtually all TEMs. Photographic emulsions are suspensions of silver halide grains in a gel. Electrons strike the halide, ionize it, and transform it to silver. The emulsions are usually supported on a polymer film or (very rarely) glass plates. Unlike polymer film, glass plates do not outgas and do not shrink during prepumping or processing. However, glass plates are heavy and occupy an enormous volume compared to polymer film, so you can't load as many into the microscope at once and you need more storage space in general. Also, Murphy's law means that your best plates will invariably break because you spend more time looking at those than the others. Most microscopists use film rather than plate, but we still sometimes call them "plates."

You do have a choice of photographic emulsion, just as you do for your own camera. Different speed emulsions are available, with the usual compromise that faster film means a larger grain size and therefore less resolution.

- In principle, for the highest-resolution images, the slowest (finest grain) film is best.
- In practice, we usually minimize the exposure time and go for the fastest film.

We usually want to minimize beam damage and blurring due to movement (drift) of the specimen/stage, so we keep the exposure short. In fact the faster film grain sizes are about 5 μm compared to about 4 μm for the slowest film, so we don't lose much resolution. The loss of resolution is more than offset by the shorter exposure times, which mean that the overall dose to the specimen is minimized. The only time you may need to use the slower film is if you have a problem with poor image contrast. This problem is more common when imaging amorphous, biological, or polymer specimens.

Although the grain size of the emulsion may be as small as a few micrometers, the actual resolution of the recorded image is worse than this because of electron spreading in the emulsion. The practical resolution may only be about 20–50 μm. Despite this degradation we still have more than 10^7 picture elements or pixels available to store information in a 10 cm × 10 cm image (Kodachrome film has the silver halide equivalent of 1.8×10^7 pixels). Film has a high DQE, although its dynamic range is rather limited. What this means is that you can easily saturate the film (change all the halide to silver). CCDs will eventually replace the photographic plate because the device size is already comparable with silver halide grains. As we've already noted, CCDs have high dynamic range. In fact, the latest CCDs boast $> 2.5 \times 10^7$ pixels, and while the cost is currently prohibitive, it is falling as is the cost of storing the data.

We use instant film for recording CRT images in STEM mode. Different film speeds are available. For example, Polaroid instant film comes in positive–negative form (Type 55) or just a positive print (Type 52), which is also slightly faster. Instant color is also available for recording false-color images from a digital CRT display (such as X-ray maps, as we discuss in Chapter 36). The main drawback to instant film is its expense. Although you can see the image "instantly," you still need to coat the print chemically to prevent degradation of the image with time.

7.5.B. Other Image Recording Methods

Digital images can be stored and retrieved magnetically, for example on floppy disks and hard disks or optically on compact disks. These devices are cheaper and easier to use than photographic recording and images on an optical disk will not degrade with time even after years of storage. To present a stored image for publication you still have to print it in some way and photographic methods still dominate. However, alternative devices such as thermal printers, laser printers, and image plates are approaching the quality required for published images.

So currently the photographic method still dominates both in TEM analog recording and hard copy output from STEM or any digital imaging system. If we can remove photographic film from the TEM it will be a major improvement, because the absorbed water degrades the vacuum.

> A photographic emulsion on a polymer support is one of the worst things you can put into a high vacuum instrument.

Both the emulsion and the support outgas, which is a major contribution to the residual pressure of hydrocarbons and water vapor in the instrument, and in turn causes contamination of the specimen. We are often caught in a compromise between drying the film before putting it in the TEM (prepumping) and avoiding the static electricity discharge which can cause white lines on your film if it is too dry. Anything that gives you a high-quality print from digital storage and doesn't require instant film recording from a CRT display will also make the microscopist's life considerably easier and microscopy cheaper.

7.6. COMPARISON OF SCANNING IMAGES AND STATIC TEM IMAGES

We have a choice of creating analog static images in conventional TEM mode or digital scanning images via electronic detection and display. Which is best? While we can only form BSE and SE images in a scanning mode, the answer is not clear for the conventional BF and DF images, and the answer depends somewhat on the contrast mechanism that is operating in the specimen, as we'll see in Chapter 22. Regardless of which detector you use, scanning images are always displayed on a CRT, and this limits the amount of information in the image. Typically, the viewing CRT will have up to 10^3 lines with a maximum of 10^3 pixels per line, giving a total of 10^6 pixels in each frame. The recording CRT may have up to 2×10^3 lines, giving a total of 2×10^6 pixels in the 10 cm $\times$ 10 cm final image. In contrast, as we just noted, a TEM image recorded directly onto photographic emulsion will have a higher information density, with more than 10^7 pixels of information available in the same 10 cm $\times$ 10 cm image. Furthermore, if a scanning image is to be recorded in a reasonable time, the electron beam can only stay on each point in the image (i.e., each pixel on the CRT) for a very short time. Typical dwell times per pixel are $\ll$1 ms and this means that the signal-to-noise ratio in a scanning image is liable to be quite low. The combination of the lower pixel density compared to a photographic emulsion and the short dwell time means that, almost invariably, STEM images are poorer in quality than static TEM images. Only in FEG STEMs does the picture quality compare with analog TEM images.

CHAPTER SUMMARY

The TEM is still in the age of analog images. We look at fluorescent screens and CRTs and we record our pictures on photographic film. However, the whole area of electron detection is in a state of rapid flux as electronic systems develop. Semiconductor detectors, scintillators, and CCDs all bring with them the advantage of digital signal collection and therefore the images can be processed and subsequently stored either magnetically or optically. Anything we say about this technology will probably be obsolete before it is published. It is probably safe to speculate that most analog detection, recording, and storage of images and diffraction patterns will eventually be replaced by digital methods.

REFERENCES

General References

Chapman, J.N., Craven, A.J., and Scott, C.P. (1989) *Ultramicroscopy* **28**, 108.

Knoll, G.F. (1989) *Radiation Detection and Measurement,* John Wiley and Sons, New York.

Nicholson, W.A.P. (1981) *J. Microsc.* **121**, 141.

Reimer, L. (1985) *Scanning Electron Microscopy,* Springer-Verlag, New York.

Yang, E.S. (1978) *Fundamentals of Semiconductor Devices,* McGraw-Hill, New York.

Pumps and Holders

8

CHAPTER PREVIEW

In the past three chapters we've described the sources, lenses, and detectors that make up a TEM. The only other parts of the instrument you need to know about in detail are the parts that, if you are not careful, can seriously degrade the quality of the information you generate. These two parts are the holder in which you put your specimen and the vacuum that surrounds it. While there isn't much you can do to improve the vacuum, beyond buying a better microscope, there is a lot you can do that will degrade the quality of the vacuum in the column and, in doing so, contaminate your specimen. So we'll tell you a few basics about how the vacuum pumps work, and how the vacuum system is put together. Although the vacuum system is under computer control in most TEMs, you still affect the vacuum by what you put in the microscope. Consequently, you need to know what not to do on those occasions that you can degrade the vacuum.

The vacuum in the stage of a typical TEM is ~10^{-5} Pa, which compares with atmospheric pressure of ~10^{5} Pa. It is quite remarkable that we can transfer a specimen into the TEM, reducing the ambient pressure at its surface by 10 orders of magnitude in a matter of a few seconds. This rapid transfer is a testament to the skills of TEM designers, and particularly the construction of the specimen holder and the airlock system. Specimen holders are the physical contact between you and your specimen across this extraordinary vacuum range. Through the holder you have to transmit all the experimental variables that you want to inflict on your specimen. The most basic requirement is that you should be able to move the specimen laterally to look at different areas. In addition we'll describe how you can tilt, rotate, heat, cool, strain, and bias the materials that you are studying. In addition to transferring useful external variables to the specimen, the holder also transmits vibrations, drift, and contamination to the specimen and may be a source of X-rays that can confuse any microanalysis that you want to perform. Care of your specimen holders is extremely important since damaged or worn holders reduce the quality of the data generated by the microscope. If you are not careful, a $10,000 holder can easily limit the information generated by a million dollar TEM.

Pumps and Holders

8

8.1. VACUUMS

You know already that electrons are strongly scattered by atoms, which accounts for the versatility of TEM, and the need for thin specimens. Strong scattering also occurs in gases and we can't send coherent, controlled electron beams through the atmosphere, so all EMs operate under vacuum. This means that your specimen has to go through an airlock into the TEM. Therefore, you can only control your specimen remotely, not directly, and this makes TEMs more expensive to build. In addition to permitting the electron beam to travel through the instrument undisturbed, the vacuum also plays a role in keeping the specimen clean (or dirty). Contamination of the specimen by vacuum-borne contaminants such as hydrocarbons and water vapor can be a problem in many aspects of TEM. Generally, the better the vacuum, the less contamination, but it is the partial pressure of contaminants, not the absolute pressure, which is important. Fortunately, the vacuum systems in most TEMs today are reasonably clean, fully automated, and their operation is transparent. Despite this, you should have some understanding of vacuums and how to control them, so this chapter will cover, very superficially, the principles of vacuum systems and pumps. For a full exposition of vacuum technology for TEMs, read Bigelow (1995) or the equally informative user's guide by O'Hanlon (1980).

First of all, a word on units—which as usual are in disarray. The SI unit of pressure is the pascal (Pa); other non-SI units are the torr and the bar. You'll come across all three units in TEM texts and in manufacturers' handbooks, so you need to know the conversions. One bar is atmospheric pressure (~760 Torr) and is equivalent to ~10^5 Pa and so

> One Torr is ~ 130 Pa, 1 Pa is 7.5×10^{-3} Torr.

We'll use the Pa, but since the Torr is still very common terminology, we'll occasionally put highly approximate torr values in parentheses to remind you of the conversion. Since we deal with very low pressures, the numbers are small, although we perversely use the expression "high vacuum" for these low pressures. We can divide vacuums into rough, low, high, and ultrahigh. Pressures between 100 and 0.1 Pa (~1 and 10^{-3} Torr) are rough vacuums, 0.1–10^{-4} Pa (~10^{-3}–10^{-6} Torr) are low, 10^{-4}–10^{-7} Pa (~10^{-6}–10^{-9} Torr) are high vacuums (HV). If the pressure is <10^{-7} Pa (~10^{-9} Torr) you have an ultrahigh vacuum (UHV). These are approximate, not standardized definitions. A typical modern TEM has a pressure inside the column of ~10^{-7} Torr (1.3×10^{-5} Pa), which is in the HV range. UHV TEMs operate below 10^{-7} Pa and the gun region of an FEG TEM operates at ~10^{-9} Pa (10^{-11} Torr). To get an electron beam inside the TEM that is not scattered by the air molecules in the column, the pressure has to be < ~0.1 Pa and this was achievable with simple mechanical pumps in the early days of the instrument. But there are good reasons to operate at much lower pressures (higher vacuums), for which you need more sophisticated and more expensive apparatus.

Generally, we use one type of pump to create a rough vacuum and another type to create the higher vacuum. The TEM is kept permanently under vacuum, unless under repair or service. If you need access to the inside of the column to change specimens, electron sources, or photographic plates, you do this via an airlock system, which can be pumped separately, as we'll explain later. There are many different kinds of pumps used in TEMs, and you often have a choice when purchasing an instrument. As with most things, you get what you pay for; a clean UHV system is very expensive. We can divide pumps into roughing pumps and HV/UHV pumps, as we'll now discuss.

8.2. ROUGHING PUMPS

The most common roughing pump is a mechanical (rotary) pump in which a belt-driven, eccentrically mounted reciprocating mechanism sucks air through an inlet valve into a

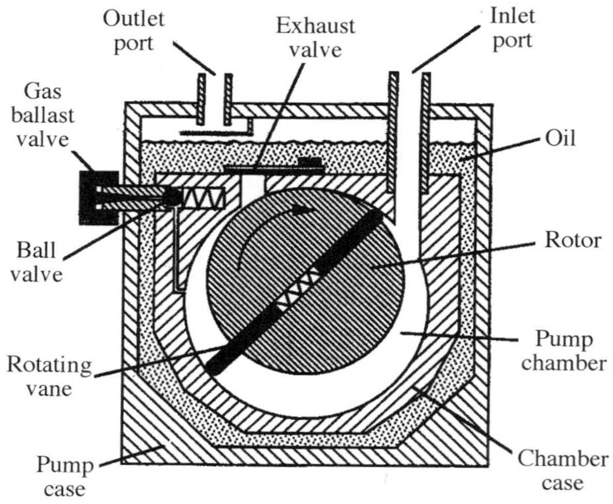

Figure 8.1. A mechanical pump for roughing vacuums. The eccentric motion of the pump creates a vacuum in the RH side when it rotates and the vacuum sucks air into the inlet valve. As the cylinder rotates further, it cuts off the inlet and forces the air through the outlet on the LH side, creating a vacuum again on the inlet side as it does so. Because of the constant contact between the rotating cylinder and the inside of the pump, oil is needed to reduce frictional heating.

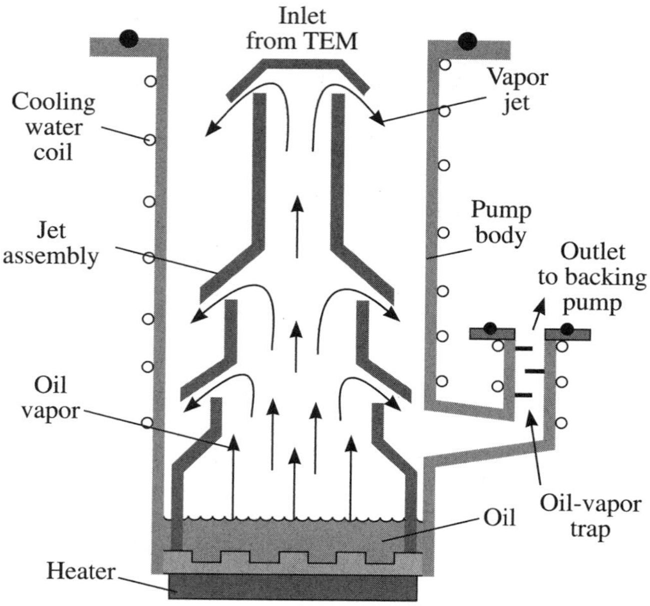

Figure 8.2. Principles of diffusion pump operation. A heater plate at the base of the pump boils synthetic oil. The expansion of the oil vapor on boiling creates a pressure, which forces the vapor up the central column and out of several holes. The stream of oil vapor pulls gas molecules out of the top of the pump down to the base, where the oil condenses and the air is pumped out of the base by a mechanical backing pump.

chamber and expels it through an exit valve, as shown in Figure 8.1. Such pumps are very reliable, relatively inexpensive, noisy and dirty, and only lower the pressure to ~10^{-1} Pa (~10^{-3} Torr). Mechanical pumps should be housed outside your TEM room, and connected to the column through a line that doesn't transmit their vibration. These pumps use a hydrocarbon oil as a medium. If you have such a pump, the line from the pump to the vacuum should contain a foreline trap to condense out oil vapor before it is deposited in the column. Also, the exhaust line from the pump must be well trapped to prevent (possibly carcinogenic) oil vapor escaping into the room where you are working. There are alternative "dry" roughing pumps which do not use oil. These are more expensive and somewhat less reliable; they do not pump to such low pressure.

8.3. HIGH/ULTRA-HIGH VACUUM PUMPS

8.3.A. Diffusion Pumps

These pumps use a hot plate to boil oil, which then forms a series of concentric vapor jets. The jets drag air molecules out of the microscope as shown in Figure 8.2, then condense onto a cold surface, freeing the air molecules which are extracted by the mechanical pump "backing" the diffusion pump. While this may seem an inefficient way to move air, diffusion pumps can in fact transport hundreds of liters of air per second, which is quite sufficient to pump out a TEM column. With no moving parts, diffusion pumps are inexpensive and very reliable, but they need external water cooling to aid condensation of the vapor. Failure of the water cooling and burn-out of the hot plate are about the only possible causes of failure. The absence of moving parts ensures vibration-free operation. As with the mechanical pump, the oil diffusion pump would contaminate the vacuum in the TEM if oil vapor were to escape into the column. To minimize this you must use synthetic nonhydrocarbon oils with low vapor pressures, such as Fomblin™ or Santovac™. A liquid-N_2 cold trap sits on top of the pump and condenses out any residual oil molecules. If you have diffusion pumps you must keep the cold traps full of liquid N_2 to maintain a clean system.

Diffusion pumps are capable of very efficient pumping from ~10^{-1} to ~10^{-9} Pa (10^{-11} Torr) and, if properly trapped, will provide a clean UHV system that is very reliable. The VG series UHV DSTEMs use only oil diffusion pumps to attain UHV conditions.

8.3.B. Turbomolecular Pumps

Turbomolecular pumps, or turbopumps, as the name implies, use a turbine to force gases from the microscope.

Figure 8.3. A turbopump (with and without its casing) which is nothing more than a small turbine that rotates at high speed. Like a jet turbine it pulls air in at the front end and forces it out of the back. The blades are designed like airfoils to enhance the flow of gas through the system.

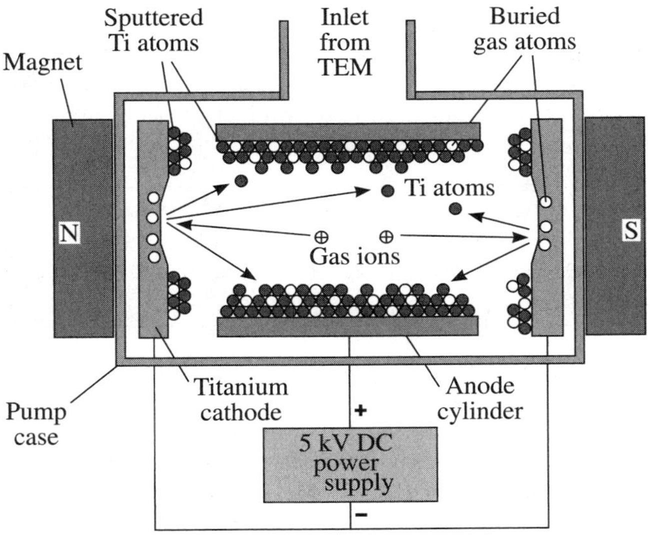

Figure 8.4. Schematic diagram showing how ion pumps trap ionized gas atoms by layers of Ti atoms at electrodes. Once trapped, the ions cannot escape until the pump is turned off.

They have many parts moving at high speeds (in excess of 20,000–50,000 rpm is common), so they are more liable to fail than diffusion pumps. The mechanics of the pump are very simple, as you can appreciate from Figure 8.3. They do not use oil so they don't introduce hydrocarbons to contaminate the microscope, and the best models (unlike earlier versions) are very quiet and almost vibration-free. In fact, modern turbopumps are being used to prepump the specimen chamber when this is critical, as in the cryotransfer technique (see Section 8.10). If you buy a turbopump, make sure to specify that its use will not transmit vibrations to the TEM column, where they would destroy the image resolution. The turbopump can start (slowly) at ambient pressures, increasing speed as the pressure is lowered, ultimately providing UHV conditions at high enough speeds. It is usual, however, to back the turbopump with a dry mechanical pump.

> Mechanical, diffusion, and turbopumps are all exhaust pumps; they pull in air from one end and expel it from the other.

8.3.C. Ion Pumps

Ion pumps do not contain oil, so they cannot contaminate the TEM column. They also have no moving parts, relying solely on the ionization process to remove air. The ion pump emits electrons from a cathode, and these spiral in a magnetic field (see Section 6.3) and ionize air molecules, which are then attracted to the cathode. The energetic gas ions sputter Ti atoms from the cathode and they condense throughout the system, mainly on the cylindrical anode, trapping gas atoms. Thus ion pumps remove gas atoms in two ways; by chemisorption on the anode surfaces and by electrical attraction to the cathodes. The smaller the ion current between the electrodes, the lower the vacuum, so the pump acts as its own vacuum gauge. Ion pumps are only efficient at high vacuums, so they are usually switched on after a diffusion pump has lowered the pressure to $< \sim 10^{-3}$ Pa (10^{-5} Torr). It is common to add ion pumps directly to the stage or gun chambers of TEMs to focus their pumping action on these important regions. Since these pumps are very common on TEMs, we include a diagram (Figure 8.4) showing how they operate.

8.3.D. Cryogenic (Adsorption) Pumps

As the name implies, cryogenic pumps (cryopumps) rely on liquid N_2 to cool molecular sieves with large surface areas. The cold surface efficiently removes air molecules from ambient pressure down to $\sim 10^{-4}$ Pa (10^{-6} Torr). Because they are oil-free, cryopumps are also used to back out ion pumps and prevent their accidental contamination through backstreaming from oil-bearing pumps.

We also use cold surfaces to enhance vacuums in the stage of most non-UHV TEMs. Such "cold fingers" or "anticontaminators" provide an alternative site (to your specimen) for condensation of residual components in the vacuum.

> Ion pumps and cryopumps are trapping pumps. They keep the air molecules within them and release them when turned off or warmed up, respectively.

The same is true if the anticontaminator in your stage is allowed to warm up; then it will degrade the vacuum around your specimen. So you must use another pump, such as a diffusion or mechanical pump, to remove the air molecules as they are released from captivity. Otherwise, this outgassing will degrade the quality of the vacuum around your specimen, increasing contamination.

8.4. THE WHOLE SYSTEM

As shown schematically in Figure 8.5 the TEM has separate pumping systems: one that evacuates the column and one that pumps the camera and screen chamber. We pump the camera separately because the film is one of the primary causes of vacuum degradation since outgassing occurs from the emulsion that contains the AgI grains. So this part of the TEM is usually pumped by a combination mechanical/diffusion pump. The stage is often pumped by a separate ion pump, turbopump, or cryopump, or some combination of these. If the instrument has an FEG, then there is a separate UHV pumping system for the gun region, which often consists of several ion pumps. Each part of the vacuum system consists of roughing pumps (mechanical or turbo) that pump out the appropriate part of the microscope to a pressure below which the HV/UHV pumps can operate.

Looking at Figure 8.5, there are three valves, which are now all computer-controlled:

- #1 connects the mechanical pump to the column (the roughing valve).
- #2 connects the mechanical pump to the bottom of the diffusion pump (the backing valve).
- #3 connects the diffusion pump directly to the TEM column (the butterfly valve).

If you're pumping down from atmospheric pressure, you first use the mechanical pump to back out the diffusion pump, till it gets to a low enough pressure so its heater can be safely switched on without oxidizing. So close #1, open #2, and close #3.

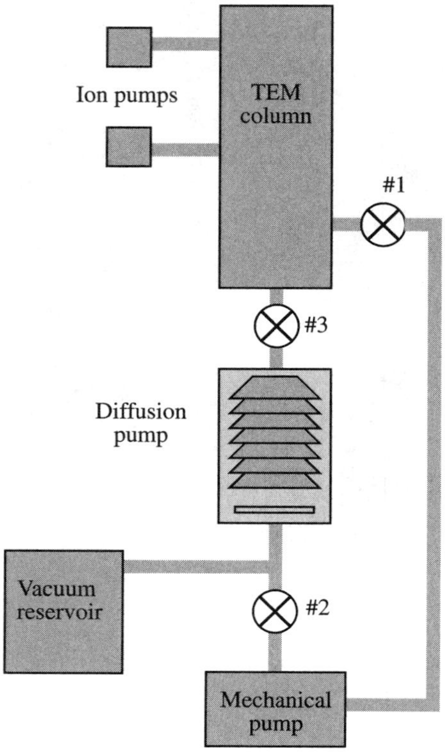

Figure 8.5. The principles of the TEM vacuum system. Often, the console display on the TEM will show a similar diagram. The mechanical pump can pump the column directly or back out the diffusion pump, which is connected directly to the base of the microscope. Ion pumps are often interfaced directly to the stage and gun areas. Computer-controlled valves separate the pumps from the column and from each other.

When the diffusion pump is warmed up, you rough out the column: open #1, close #2 and #3, until the column is at a low enough pressure that the diffusion pump can be used.

At this point, close #1, open #2 and then #3, so the diffusion pump is open to the TEM and also continuously backed by the mechanical pump. Alternatively, a vacuum reservoir is attached between the mechanical and diffusion pumps. When the reservoir is pumped to < 0.1 Pa, the mechanical pump is closed off and the diffusion pump exhausts into the reservoir. When the pressure builds in the reservoir, the mechanical pump will automatically switch on and lower the pressure.

Similar arrangements work for other pumps; e.g., the diffusion pump is used to lower the pressure in the stage and gun sufficiently for the ion pumps to be switched on, and so on. In most TEMs the stage and gun have significantly better vacuums than the camera region, so the camera/screen is isolated from the rest of the column by a differential pumping aperture (not shown in Figure 8.5). This aperture often coincides with the BFP of the projector lens, since all the electrons have got to pass through it and the diffraction pattern in the BFP localizes all the electron trajectories close to the optic axis. A similar arrangement exists between the stage and gun in FEG systems to preserve the tip in case of a vacuum leak in the stage.

> The advent of high-quality digital recording which will remove the need for film in the camera will do more to improve the quality of vacuums in TEMs than any advances in pumping technology.

8.5. LEAK DETECTION

Nature abhors a vacuum, as Francois Rabelais said in 1534, and that's the reason why the pumps have to keep pumping: the TEM leaks. But some leaks are too large for the pumps to handle, and then the instrument performance degrades. For example, you might not be able to run the electron gun, so your TEM is useless. Under these circumstances, you have to find the leak, cure it, and repump out the instrument (this is usually a job for your service engineer). Leak detection involves using a mass spectrometer, which can be put into the pumping lines of the microscope. You then release helium gas close to the various parts of the TEM where you suspect a leak (e.g., the stage airlock, which sees a lot of use, is a common point of failure). The small helium atoms are relatively easily sucked into the column through any leak and register on the mass spectrometer. When a leak is isolated, the TEM may have to be opened to the atmosphere to permit replacement of the defective part, such as the O-ring seals.

The most common cause of a leak is your specimen holder. The O-ring seal on the shaft of a side-entry holder (see the second half of this chapter) is easily contaminated with dust or a hair since it is continually inserted and extracted from the column, and left on the bench while you pore over it. Never touch the O-ring, make sure it doesn't dry out, but if it does, lubricate it with a very thin film of vacuum grease.

After repairing a leak, when you've pumped down again, it is often useful to "bake" the column. Baking means heating the internal surfaces to >100°C (or >150–200°C in UHV TEMs) to boil off residual water vapor and hydrocarbons that may have entered the system when it was down to air. Usually, you can bake by leaving the lenses running without their cooling water (check this *very* carefully with the manufacturer before proceeding). In some cases, special heating panels are constructed around the column. Baking can also introduce other leaks as the whole system expands and then contracts, so sometimes leak detection and cure is an iterative process. For UHV systems, you *must* bake to reach the ultimate vacuum, and the higher the temperature the better.

> Be wary, however, since sometimes the TEM accessories, such as XEDS and EELS systems, are not designed to be baked to the same high temperature as the column.

8.6. CONTAMINATION: HYDROCARBONS AND WATER VAPOR

As we said right at the start of the chapter, the vacuum can be a source of contamination, particularly residual hydrocarbons from the pump oil which crack under the electron beam. Carbonaceous material then deposits on your carefully thinned specimen, making it difficult to do sensible high-resolution imaging or microanalysis. So a clean vacuum (one in which the hydrocarbon partial pressure is $< 10^{-9}$ Pa) is essential. Fortunately, most modern TEMs are relatively contamination-free, particularly if you use appropriate traps on the pumps and synthetic oils.

However, even if you've paid dearly for a clean vacuum system, contamination often occurs and it comes primarily through the airlock with your specimen. You can minimize this by heating the specimen to >100°C in a heating holder or with a halogen lamp in the prepump chamber, or cooling the specimen to liquid-N_2 temperatures in a cooling holder. It may help if the prepump chamber is pumped with an oil-free pump. More recently, plasma ashing of the specimen holder and specimen prior to insertion in the TEM has proven a very successful way to ensure a clean specimen, but this is expensive. Polymers and biological specimens can easily introduce hydrocarbon contaminants, as they outgas in the vacuum, so it is sensible to cool such specimens (since heating or plasma ashing destroys them). However, when you cool your specimen, it attracts water vapor which condenses as ice on the surface; so load your specimen first, then cool it down in the TEM before you switch on the beam. A low partial pressure of H_2O in the vacuum is obviously essential. Also, warm up any cooled specimens in the TEM before bringing them out to ambient atmosphere, otherwise they will immediately ice up (unless it's a *very* dry winter's day in Minnesota).

There will be more about this in the ensuing sections on specimen holders.

In addition to the specimen, you personally can be a major source of contamination unless you take care never to touch anything that will enter the vacuum, i.e., the specimen itself, the grids, specimen holder (beyond the O-ring seal on the rod), clamping rings, replacement diaphragms, new filaments, replacement Wehnelt, components of XEDS and EELS systems, etc. Use latex gloves whenever you load a specimen, and don't breathe on it. Store specimen holders and specimens in a dry box containing a desiccant such as silica gel, which should be replaced regularly. *Always* prepump fresh film in a vacuum desiccator (which is sometimes integrated into the TEM itself). Simple precautions like this will minimize contamination of your specimens and the microscope in general and bring a much greater return in terms of good data per TEM session.

8.7. SPECIMEN HOLDERS AND STAGES

To look at your specimen, you place it in a specimen holder and insert this assembly into the TEM stage. Therefore, there are two key components which are often not separated, namely, the holder and the stage. In this part of the chapter, we will emphasize the holder but the stage is also critical. Suitable design of the stage is the essential precursor to computer-controlled TEM, which is already appearing.

The cold trap, cold finger, or cryoblades are a critical part of the stage. Ideally, this cold finger will completely surround the specimen. However, the cold surfaces, usually brass, provide a source of stray electrons and X-rays which is undesirable for AEM (see Chapter 33), so these blades should be removable.

X-ray diffractometers use goniometers to hold and tilt the specimen; so do TEMs. Conventional SEMs use a stub on which you mount the specimen so that you can bring the specimen close to the objective lens. However, the new high-resolution SEMs use a specimen holder which is very similar to those used in the TEM, because the specimen is inserted inside the lens, rather than underneath and outside it.

The reason the specimen holder is so important in TEM is that your specimen must invariably be located within the objective lens and the aberrations associated with the objective lens determine the resolution of the TEM.

Historically, microscopists have used two different designs and a lot of what you'll read has a strong historical background.

- The traditional side-entry holder is a rod with a motor attached to tilt and/or rotate the specimen and a lead connecting it to a power supply and control box, or liquid-N_2 dewar.
- The traditional top-entry holder is a cartridge which you load into the TEM but is detached from the outside world when you use the microscope.

The actual cup that holds your specimen is either 2.3 mm or 3.05 mm in diameter, so the specimen disk or support grid has to be the same dimension, as we'll see in Chapter 10. The reasons for these dimensions are again partly historical. In the top-entry holder the specimen and part of the holder fit through the bore of the upper polepiece (see Chapter 6). Clearly, the specimen must be smaller than the bore diameter. So the original top-entry holders used small specimens.

Side-entry holders are more versatile and larger specimen dimensions first appeared when they were introduced. However, side-entry holders connect the specimen directly to the outside world via a long lever arm, which is undesirable, unstable, and also not necessary in many cases! Ideally, the side-entry holder should leave the specimen in the stage, not connected to the outside world, and all manipulations should be conducted through the stage itself, not the holder. This ideal is being approached as stages become more computer-controlled.

8.8. SIDE-ENTRY HOLDERS

Side-entry holders are now the standard, although their design has changed quite radically. The traditional design is shown in Figure 8.6. The key parts of the holder are:

- *The O-ring,* which is one mechanical link to the microscope column. Some holders have two O-rings and the gap between the O-rings is pumped separately to improve the vacuum.
- *The jewel bearing,* which is the other mechanical link to the microscope column. You push on this bearing to move your specimen back and forth and from side to side. Like the O-ring, you must keep the bearing clean otherwise the specimen will not be stable.
- *The cup,* which actually holds your specimen and thus provides the immediate environment which is seen by stray electrons and any X-rays coming down the column. So cups in holders for AEM are made of Be to minimize the gener-

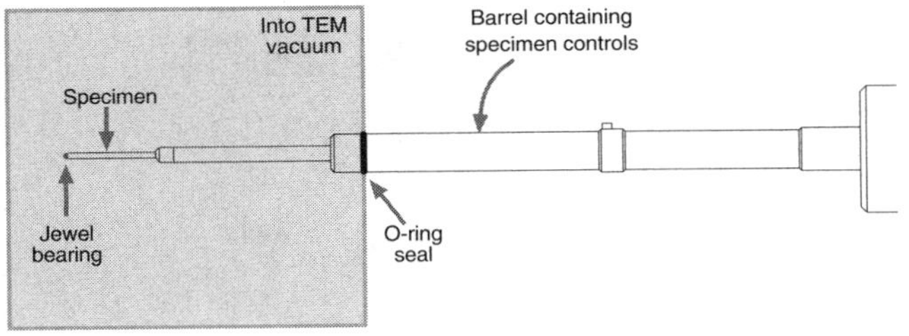

Figure 8.6. Principal parts of a side-entry holder that is held in the goniometer stage. The specimen is clamped into the cup at the end of the rod. A small jewel at the end of the rod (usually sapphire) fits into another jewel bearing in the stage to provide a stable base for manipulating the specimen. The O-ring seals the end of the holder inside the vacuum. Manipulating the specmen is accomplished from outside the column via controls within the rod.

ation of X-rays that would interfere with microanalysis.

- *The clamping ring or screw,* which holds the specimen in the cup. This ring, which may also be Be, must be carefully designed. It must hold your specimen firmly (so, e.g., magnetic disks cannot be pulled out of the cup by the lens field). However, the ring must not be so difficult to tighten that you put undue pressure on your specimen, since brittle disks may break as you are loading them. There are two kinds: screw-thread rings, which are easier to control and do not damage metals, but you'll find they may break ceramics because they transfer shear stresses to the disk; spring clips are difficult for the novice to master, but with practice you'll find they offer more control over the load that you put on the specimen, so we recommend them for the experienced ceramist. Unfortunately, no one makes Be spring clips!

8.9. TOP-ENTRY HOLDERS

Top-entry holders are becoming less common because they essentially preclude XEDS analysis in the TEM. Also, it is more difficult to design such holders so that the specimen can be manipulated (e.g., rotated or strained). Their great advantage was that they were much less susceptible to drift since they were not connected directly to the outside, so early HRTEM required top-entry holders. Today, however, all TEMS up to 400 kV use side-entry holders.

Another drawback of such holders is that the bore of the objective lens must be asymmetric, which actually limits the ultimate resolution by constraining the lens designer. Figure 8.7 shows a schematic diagram of such a holder.

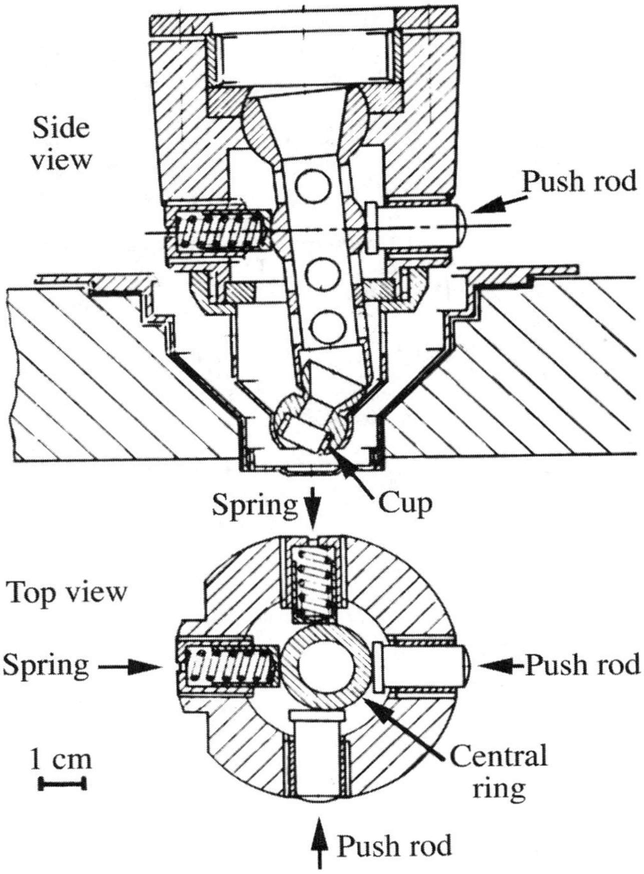

Figure 8.7. Cross section of a top-entry holder. The cartridge has a cone shape which fits into the tapered bore of the objective lens polepiece. The specimen sits in a cup at the base of a column through the cone down which the incident beam travels. Simple manipulations such as tilting or rotating require complex micromechanical design, since the specimen is at the base of the cartridge and completely surrounded by the polepiece. To tilt, e.g., as shown in the upper diagram, push rods are pressed against springs in two orthogonal directions, displacing a ring around the column (lower diagram), thus tilting the specimen cup.

8.10. DIFFERENT TYPES OF HOLDERS

One feature of TEMs which may surprise you if you are a new user is the wide variety of holders which is available. Figure 8.8 shows illustrations of some different designs for the side-entry holder:

- *Single-tilt holder:* This is the basic holder with which any novice should start practicing. You can only tilt around the axis of the rod. It is relatively cheap, robust, and can at least give you some idea of the usefulness of tilting a specimen for diffraction contrast studies.

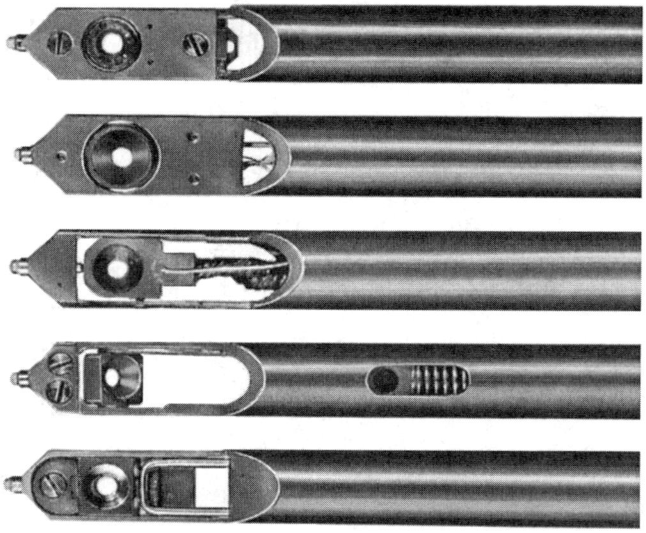

Figure 8.8. Examples of different designs for the side-entry holder. From the top, they are: a rotation holder, a heating holder, a cooling holder, a double-tilt holder, and a single-tilt holder.

- *Quick change holder:* This is also a single-tilt holder that clamps the specimen with a lever arm which you raise and lower onto your disk or grid. It doesn't put a high stress on the specimen, but it doesn't hold it very strongly either. Don't use it for magnetic specimens, but it's great for ceramics. Different retainers can be substituted for the clamp shown in Figure 8.8 (bottom), creating a more versatile multipurpose holder.
- *Multiple-specimen holder:* This is usually a single-tilt holder, but you can load up to five specimens into the column at one time, as shown in Figure 8.9A. A two-specimen, double-tilt version is also available (Figure 8.9B). Such holders can be useful if you are not very good at specimen preparation, or you want to compare different specimens under identical conditions without turning off the beam. However, in modern TEMs, specimen exchange is relatively quick, except in UHV instruments where the multiholder is probably more valuable.
- *Bulk specimen holder:* This holder is used for surface imaging and diffraction, e.g., using SE or BSE in a STEM or for reflection diffraction and imaging in a TEM (see Chapter 31 for more about these techniques). The bulk specimen is larger than the traditional 3-mm disk (usually ~10 mm x 5 mm) so if you can create a thin specimen of these dimensions, the bulk holder

A

B

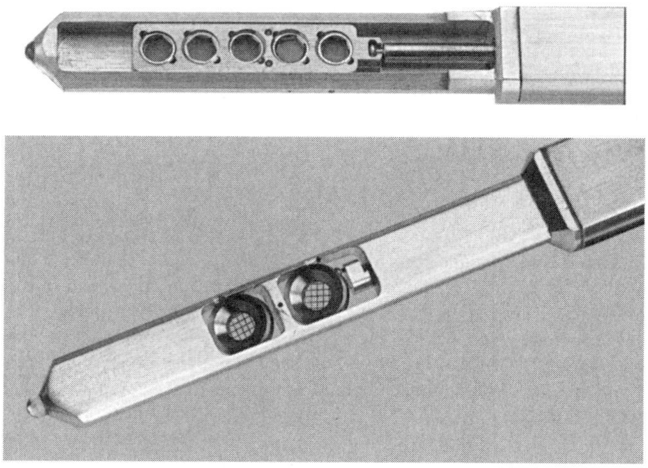

Figure 8.9. Multiple-specimen holders. (A) Five-specimen single-tilt and (B) two-specimen double-tilt.

will allow you to sample more of your material at one time (Figure 8.10).

> So don't always think that you are limited to 3-mm specimens!

- *Double-tilt holder:* This is the most popular holder since it gives you the most flexibility in orienting the specimen. It is absolutely essential for imaging and diffraction studies of crystalline specimens. The tilt axes are fixed as two orthogonal directions. In some designs, you can

Figure 8.10. A bulk holder for large specimens.

remove the cup while the specimen is in place, which means that you can reinsert your specimen in the same orientation. This feature is extremely useful if your specimen is robust.

- *Tilt-rotate holder:* You would often like to be able to select the tilt axis. This holder lets you do just that and is particularly advantageous for the side-entry holder, since the tilt axis is then always parallel to the rod of the holder which also gives the largest tilt angle.
- *Low-background holder:* The cup and clamping ring are made of Be to minimize the generation of bremsstrahlung X-rays and characteristic X-rays. So they are required for XEDS studies. They can be double or single tilt and may be cooled also.
- *Heating holder:* Such holders in a conventional TEM can go to ~1300°C, which is measured by a thermocouple attached to the cup. In HVEMs, the temperature can go higher because of the larger gap between the polepieces. You have to be careful to calibrate the temperature and remember that the temperature may be different for different specimens. You should also be sure that the material you are studying does not form a eutectic alloy with the material forming the holder! If the eutectic does form, it will have a lower melting point, so you may deposit part of your specimen and the holder on the objective lens, or down onto the screen, if the microscope is well aligned.
- *Cooling holder:* This is available for either liquid-N_2 or liquid-He temperatures. These holders, which can be single or double tilt, are a great asset for XEDS, EELS, and CBED studies since they minimize surface-borne contamination. They are also essential for *in situ* studies of superconducting materials (both high and low T_c) and ideal for polymers or biological tissue. However, you should remember that the cold holder can also act as a small cryopump, so that it actually attracts contamination. Since you are necessarily changing the temperature at the specimen relative to its surroundings, be prepared for specimen drift. It takes time for the whole system to stabilize.
- *Cryo-transfer holder:* Certain specimens are prepared at cryogenic temperatures such as liquids, latex emulsions, and tissue in general. This holder permits you to transfer such cold specimens into the TEM without water vapor from the atmosphere condensing as ice on the surface.
- *Straining holder:* This holder clamps the specimen at both ends and then applies a load to one end, via a load cell or screw-thread mechanism, as shown in Figure 8.11. The sample is often in the shape of a small tensile specimen and is thinned in the middle of the gauge length (see inset). The motion of dislocations, cracks, etc. are then easily monitored, so a video camera is an essential accessory. You can vary the load to study cyclic as well as tensile loading, and the strain rate is another variable that is easily controlled. In Figure 8.11 a furnace is present, so the specimen can be heated while under load.
- *EBIC and CL holders:* The essential feature is the electrical feed-through that allows you to control the charge recombination in a semiconductor or certain mineral specimens by applying a bias across the specimen surface.

Beware: Heating and straining holders, in particular, can produce effects in thin foils that are totally uncharacteristic of your bulk specimen. So you must use these holders carefully and interpret your results cautiously. Often, surface reactions will dominate internal reactions when you are trying to induce a phase transformation by heating. The surface may also stop grain boundaries from migrating at temperatures where they would do so in the bulk material. Obviously, defect motion under applied stress may also be strongly affected since the 3D stress

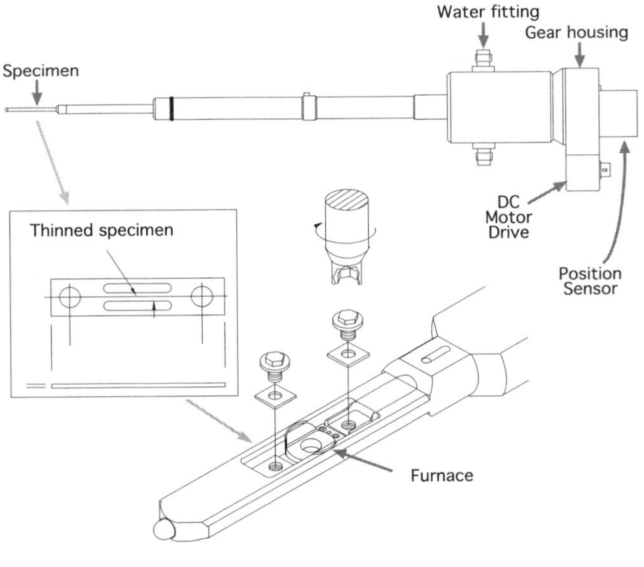

Figure 8.11. A side-entry combined straining and heating holder. The specimen looks like a miniature tensile specimen (inset) and is clamped at either end by hex screws. There is a screw-thread arrangement for pulling the specimen contained within the rod. The furnace surrounds the central thin portion of the specimen.

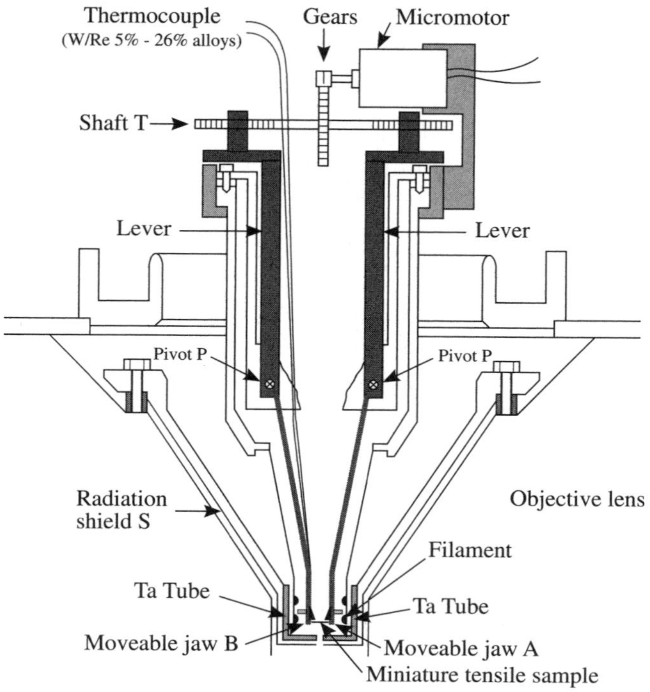

Figure 8.12. A top-entry, heating-straining holder which can be used at temperatures up to 2300 K in a 3-MV HVEM.

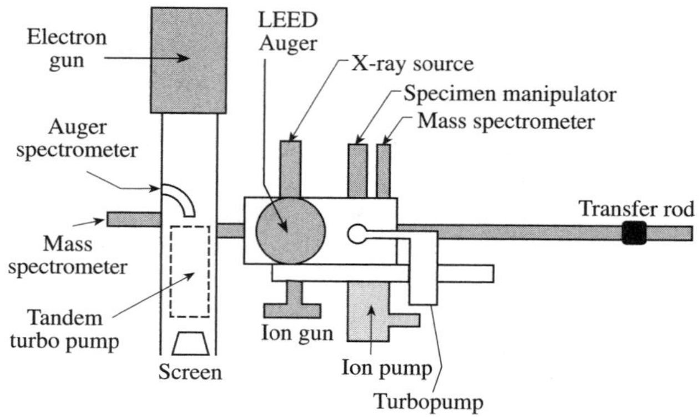

Figure 8.13. Schematic diagram showing the Hitachi H9000 UHV TEM. This instrument is equipped with a prechamber with LEED, Auger, and an ion gun which can be used to clean the specimen, allowing UHV surface analysis to be carried out on the TEM specimen. The holder has to transfer the specimen through a prepump chamber where it is ion-cleaned before going into the column.

field will be very different in bulk specimens compared to thin foils.

These problems can be overcome to some extent if you use thicker specimens and examine them in an HVEM, or at least an IVEM, and the whole field of *in situ* studies, particularly heating and/or straining, is best performed in such microscopes (Butler and Hale 1981 and Section 31.12). However, the high-energy electrons in these microscopes may introduce lattice defects that affect the very phenomenon that you want to study, e.g., beam-induced vacancies can change diffusional phase transformation kinetics very easily.

It is also possible, but much more difficult and expensive, to manipulate specimens in top-entry stages. The top-entry holder shown in Figure 8.12 is a heating-straining holder which is reported to be capable of operating at temperatures up to 2300 K (Komatsu *et al.* 1994). The heat is provided by a coaxial Ta tube which supports the W heater filament, as shown in the figure. The holder is used in a 3-MV microscope where the specimen diameter is 5 mm. The larger specimen diameter means that the disk can be shaped as a small tensile specimen and still be quite robust.

There are also special combinations of holders and stages which have been optimized for particular applications. The example shown in Figure 8.13 has been optimized to combine surface studies using low-energy electron diffraction (LEED) and Auger analysis with TEM. The prechamber is fitted with an ion gun to clean the sample before the surface is analyzed. The specimen can then be moved on into the TEM column for transmission studies. A similar prechamber has been used elsewhere to provide a method to clean the sample before growing thin films on the surface by molecular-beam epitaxy (MBE) or thermal evaporation.

One of the reasons for using higher accelerating voltages is that this gives more room in the specimen-stage region. Thus even 400-kV microscopes can be fitted with a small, differentially pumped environmental chamber. Such a chamber allows *in situ* studies of corrosion, degradation of catalysts, etc., especially when combined with a heating holder.

The article by Valdrè and Goringe (1971) gives a detailed description of several TEM holders.

CHAPTER SUMMARY

The vacuum and the holder are the two parts of the TEM that most closely affect your specimen. You have to treat both carefully if you want to be sure of getting the most out of your TEM. The vacuum is usually automated, so you don't have too much control over it. However, you can degrade the vacuum easily if you are a careless operator; for example, if you don't bother to prepump your film, and if you handle the specimen holder without gloves. In fact, you should treat the specimen holder as if it were a rare jewel; it may actually contain a couple of synthetic ones and it certainly costs as much as a diamond of several carats!

With the range of holders available today, you can conduct many standard materials science experiments on your thin specimen while observing it in the TEM. However, if you're looking at crystalline material, the most common manipulation remains tilting in two orthogonal directions to orient different crystal planes parallel to the electron beam. You'll understand why this is important after you've finished reading Parts II and III.

REFERENCES

General References

Watt, I.M. (1985) *The Principles and Practice of Electron Microscopy,* Cambridge University Press, New York. See Appendix 1.

Bigelow, W.C. (1995) *Vacuum Methods in Electron Microscopy,* in the series *Practical Methods in Electron Microscopy,* **15** (Ed. A.M. Glauert), Portland Press, London. An essential reference.

O'Hanlon, J.F. (1980) *A User's Guide to Vacuum Technology,* John Wiley and Sons, New York.

Specific References

Butler, E.P. and Hale, K.F. (1981) *Dynamic Experiments in the Electron Microscope,* in the series *Practical Methods in Electron Microscopy,* **9** (Ed. A.M. Glauert), Elsevier, Amsterdam.

Komatsu, M., Mori, H., and Iwasaki, K. (1994) *J. Am. Ceram. Soc.* **77**, 839.

Valdrè, U. and Goringe, M.J. (1971) in *Electron Microscopy in Materials Science* (Ed. U. Valdrè), p. 208, Academic Press, New York.

The Instrument

9

CHAPTER PREVIEW

We've introduced all the essential components of the TEM. Now it's time to see how the guns, lenses, and detectors are combined to form the microscope. Just as we do for the visible-light microscope (VLM), it's convenient to divide the TEM up into three components: the illumination system, the objective lens/stage, and the imaging system. The illumination system comprises the gun and the condenser lenses and its role is to take the electrons from the source and transfer them to your specimen. You can operate the illumination system in two principal modes: parallel beam and convergent beam. The first mode is used for TEM imaging and diffraction, while the second is used for scanning (STEM) imaging, microanalysis, and microdiffraction.

The objective lens/stage system is the heart of the TEM. The critical region usually extends over less than 1 cm along the length of the column. Here is where all the beam–specimen interactions take place and here we create the bright-field and dark-field images and selected-area diffraction patterns (SAD) that are the fundamental TEM operations. Likewise, it is here that we manipulate a scanning beam to form STEM images and diffraction patterns. The quality of the image formed by the objective lens controls the resolution of the image that you view and record. We therefore often say that the objective lens is the most important lens. The reason is not simply because the objective lens forms the first image, but rather because the specimen must be placed so close to the center of this short-focal-length lens that it is impossible to make a perfect lens (or even a very good lens by visible-light standards).

The imaging system uses several lenses to magnify the image or the diffraction pattern produced by the objective lens and to focus these on the viewing screen. We'll refer to the magnifying lenses as the intermediate and diffraction lenses and the final lens as the projector lens (it projects an image on the viewing screen). Alternatively, an electron detector coupled to a TV/CRT can be used to display the STEM image. So all TEM operations involve observing the electrons on a viewing screen of some form, with or without a specimen in place. In many modern-day TEMs you will have a button for focus, another for magnification, and another for diffraction (or a slide on the computer screen).

The purpose of this chapter is to go through the principal functions of each of the three components and give you some feel for what is happening in the microscope when you "press the button." The more you understand the operation of the TEM, the better you can be sure that you are getting the most out of the instrument.

The Instrument

9

9.1. THE ILLUMINATION SYSTEM

The illumination system takes the electrons from the gun and transfers them to the specimen giving either a broad beam or a focused beam. We can think of these two cases as the equivalent of wide-field illumination or a spotlight. In Chapter 5 we described how the gun produces an image of the source (called a crossover). This crossover acts as the object for the illumination system which consists of two or three condenser lenses (which we'll call C1, etc.). We will discuss the two different ways to use the illumination system: we'll refer to these as forming a parallel beam (it is almost never truly parallel) or a convergent beam.

9.1.A. TEM Operation Using a Parallel Beam

In the traditional TEM mode the first two condenser lenses (C1, C2) are adjusted to illuminate the specimen with a parallel beam of electrons typically several micrometers across at reasonable magnifications (20,000×–100,000×). As shown in Figure 9.1, the C1 lens first forms a demagnified image of the gun crossover. In the case of a thermionic source, the original crossover may be several tens of micrometers across, and this is demagnified by an order of magnitude or more: in the case of an FEG, the source size may be less than the desired illumination area on the specimen so it may be necessary to magnify the crossover—the condenser lenses don't always condense! To produce a parallel beam you adjust the C2 lens to produce an underfocused image of the C1 crossover.

> Remember that the convergence angles (α) are so small that the ray diagrams are drawn with highly exaggerated angles, and while the beam in Figure 9.1A is not exactly parallel to the optic axis, α under these conditions is $<10^{-4}$ rads (0.0057°), which is effectively a parallel beam.

In TEMs used for generating the very small electron beams we need in STEM and AEM, the upper polepiece, of the objective lens is also used to control the beam hitting the specimen as shown in Figure 9.1B. Now the C2 lens is focused to produce an image (of the crossover) at the front focal plane of the upper objective polepiece, which then generates a broad parallel beam of electrons incident on the specimen.

> Consider the question: How is this argument consistent?

When the beam is parallel, it is as coherent as possible. We'll see in Chapter 18 that parallel illumination is essential to get the sharpest diffraction patterns as well as the best image contrast. (It is also usually assumed in the interpretation of our images that the beam is parallel.) Usually you underfocus C2 until the illuminated area on the specimen fills the viewing screen. A higher magnification means strengthening C2 so the beam illuminates less of the specimen (so you see, it isn't really "parallel," just not very "convergent").

In the parallel-beam TEM mode, there is usually no need to change C1, which is therefore kept at some intermediate setting, recommended by the manufacturer. The only other variable is the C2 aperture. A small aperture reduces the electron current falling on your specimen. However, if you use a smaller aperture, you decrease the angle of beam convergence and therefore increase the coherence of the beam, as is evident from Figure 9.2.

9.1.B. Convergent-Beam (S)TEM Mode

Now, there are times when you may wish to focus the beam more, so that the intensity of the beam on a specific area of the specimen is increased. Let's look at various ways to do this.

If you want to minimize the area of the specimen that you are illuminating, you simply change the C2 lens so it is

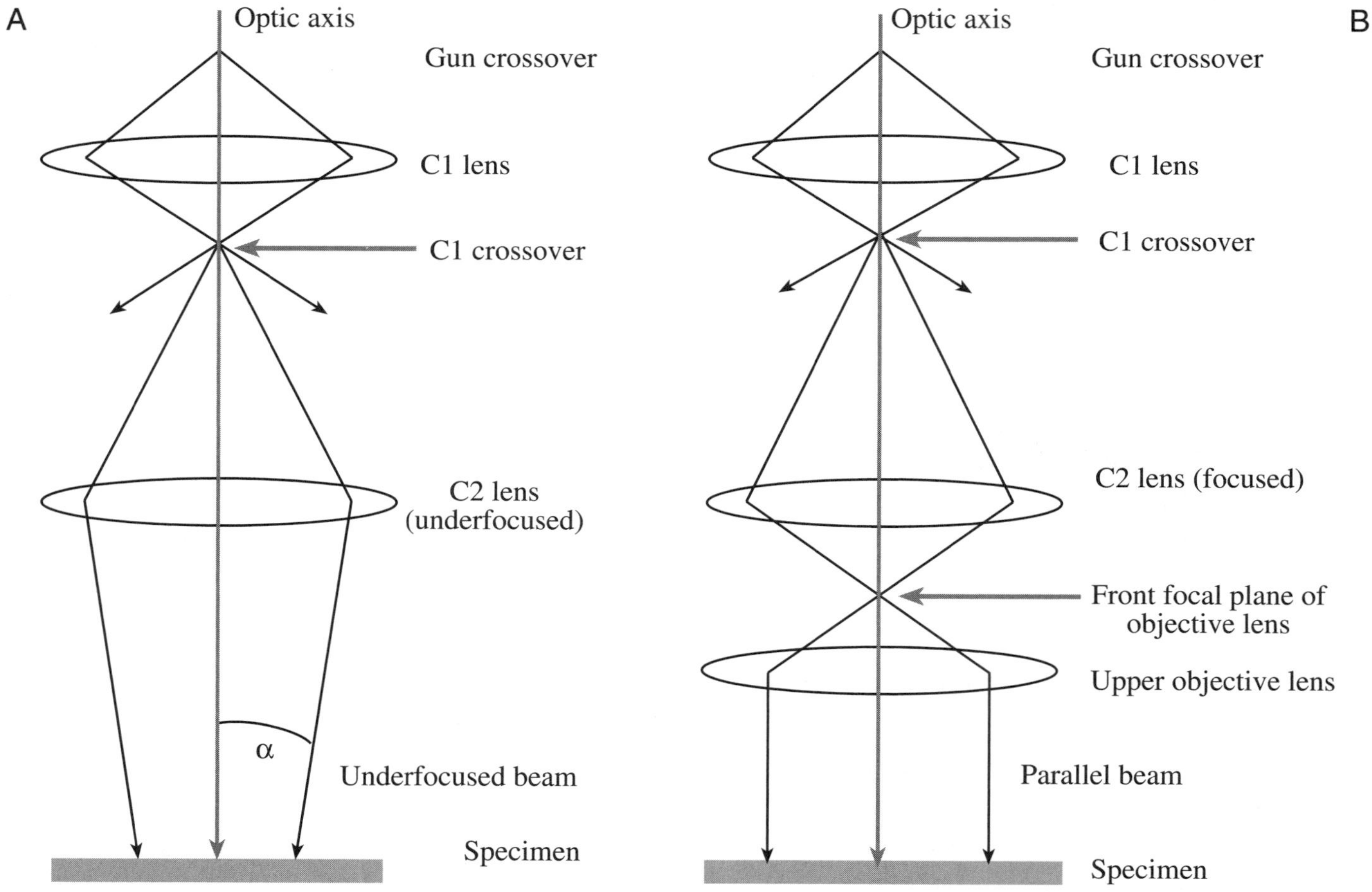

Figure 9.1. Parallel-beam operation in the TEM (A) using just the C1 and an underfocused C2 lens and (B) using the C1 and C2 lenses to image the source at the front focal plane of the upper objective lens.

focused rather than underfocused, and you form an image of the C1 crossover at the specimen, as shown in Figure 9.3. This is the condition under which you can view the source image to adjust its saturation or to measure the dimensions of the beam. When C2 is focused like this, the beam is at its least coherent and most convergent. While the intensity of illumination on the viewing screen will be greatest, your image contrast will be reduced. Ideally, for routine TEM work, your specimen should always be thin enough so that you never have to operate with C2 focused but, in practice, you'll often find yourself focusing C2 to compensate for poor transmission through a thick specimen.

There are times when we need to deliberately create a focused convergent beam at the specimen. We then use the other principal way to operate the illumination system: the convergent-beam mode. When you use this mode you won't immediately see a useful image of your specimen; the convergence destroys the coherency and the image contrast. So to see an image we have to scan the beam; this mode of operation of the illumination system is standard for STEM and AEM.

> The convergent beam is a probe. We use such a probe when we want to localize the signals coming from the specimen, as in microanalysis or convergent-beam (also known as micro or nano) diffraction.

Now, unless you have an FEG, it isn't possible to use just the C1 and C2 lenses as in Figure 9.3 to converge the beam to as small a probe as you would like (<10 nm). This is because the C1 and C2 lenses can't demagnify the gun crossover sufficiently. So the usual solution is to con-

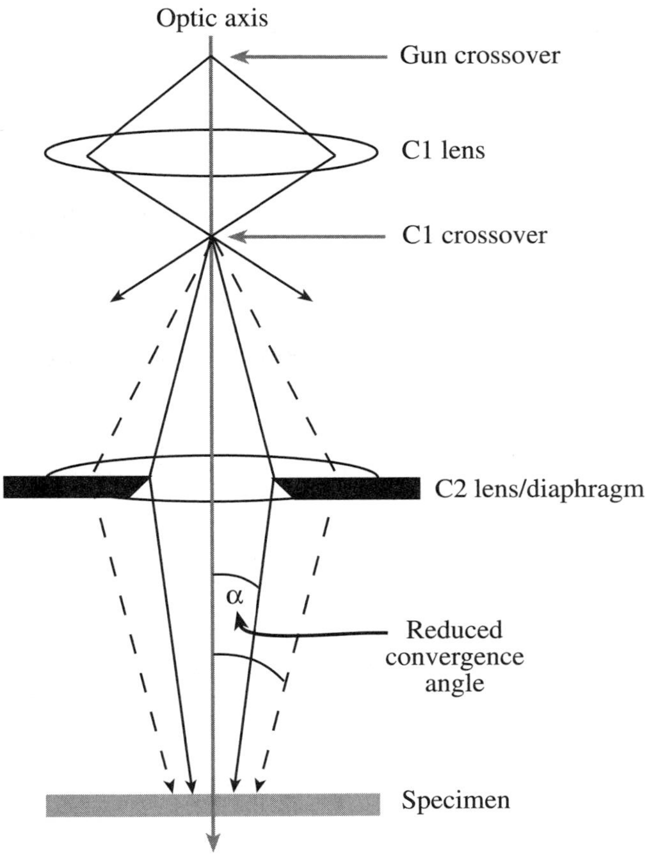

Figure 9.2. Effect of the C2 aperture on the beam coherence: a smaller aperture creates a more parallel, more coherent beam.

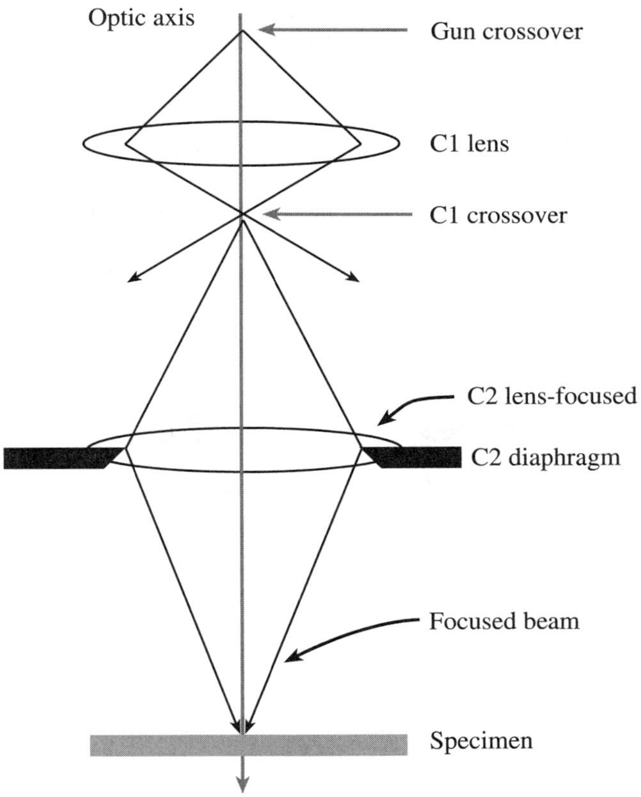

Figure 9.3. A focused C2 lens illuminates a small area of the specimen with a nonparallel beam.

vert the upper polepiece of the objective lens into a third condenser lens, which we then call a condenser-objective lens (C3). We make the upper part of the objective lens much stronger than usual and weaken C2 or turn it off, as shown in Figure 9.4. In addition, C1 must be strongly excited so the image of the gun crossover is a long way from C3. Thus the C3 image distance (v) is much less than the object distance (u), which gives a large demagnification of the C1 crossover (see equation 6.2). From Figure 9.4 you can see that although C2 is switched off, the C2 aperture still controls the convergence angle (α) of the beam on the specimen. As was the case for parallel-beam mode, a smaller C2 aperture gives a smaller α. You'll see later that the correct choice of C2 aperture is important in convergent-beam electron diffraction (CBED) and also in defining the exact dimensions of the probe for X-ray microanalysis (see Chapter 36).

The role of C1 here is fundamentally different from its role in parallel-beam TEM; now the C1 lens is used directly to form the probe because we adjust it to change the probe size at the specimen. As shown in Figure 9.5, a strong C1 gives you a small probe while a weak C1 creates a large probe. This difference occurs because increasing the strength of C1 shortens its v, thus lengthening u for the probe-forming C3 lens and therefore increasing the C3 demagnification. When convergent probe TEMs were first constructed, it was not possible to design a C3 lens that would give both a parallel and a convergent beam with the same polepiece. This problem was overcome in the mid-1970s by the introduction of an extra lens between C2 and C3 (not shown in the diagrams) and this auxiliary lens is now standard on TEMs that also operate as STEMs.

9.1.C. Translating and Tilting the Beam

There are certain operations where we need to translate the beam laterally on the specimen (e.g., to position a fine probe on a feature of interest for microanalysis). Similarly, there are times when we need to tilt the beam off axis so it impinges on the specimen at a specific angle (e.g., for centered dark-field imaging using a specific diffraction spot which we describe in Section 9.3.C). Ray diagrams to explain translating and tilting are shown in Figures 9.6A and B. Both operations are accomplished by varying the

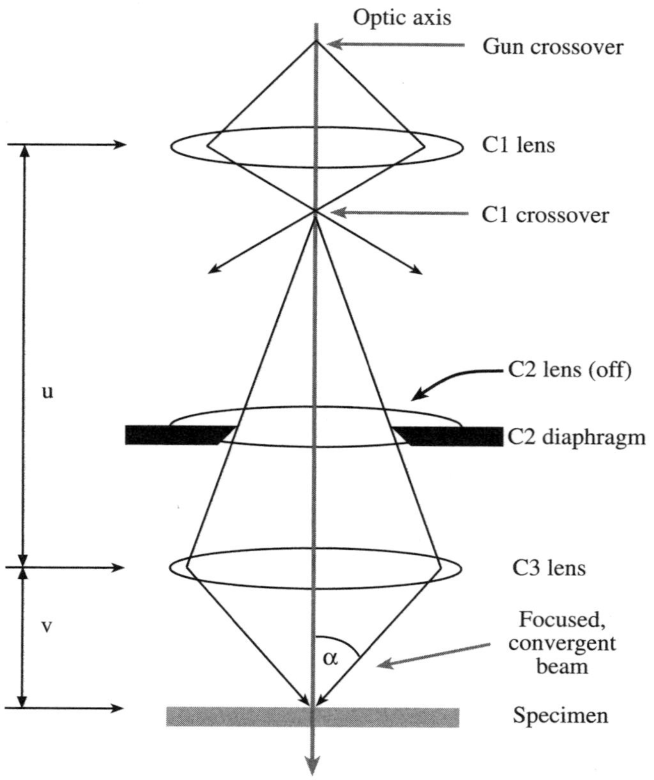

Figure 9.4. Use of the objective polepiece as a third condenser lens (also called a condenser-objective, or C3, lens) gives the smallest possible probe and large convergence angles. The large *u/v* ratio gives the maximum demagnification of the image of the gun crossover.

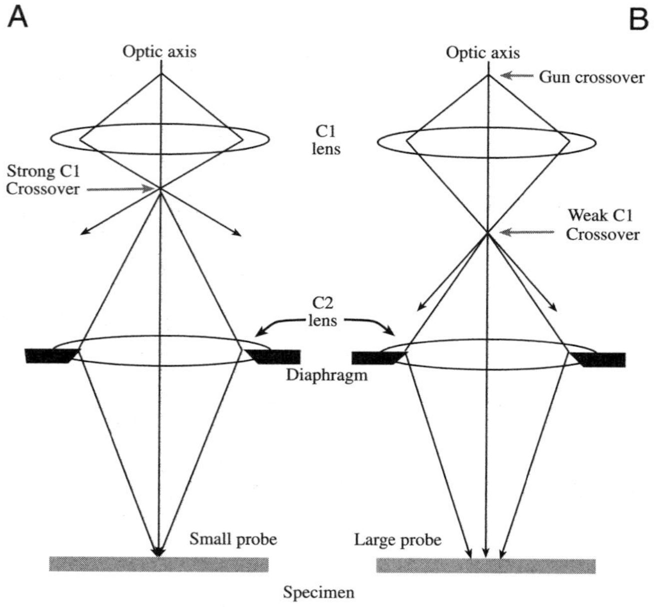

Figure 9.5. Effect of the C1 lens strength on probe size: a stronger C1 lens (A) results in greater demagnification by any subsequent lens (C2 or C3), giving a smaller electron beam at the specimen. A weaker lens (B) gives a broader probe.

current through potentiometers, which we'll call *scan coils*. We use these scan coils (of which there are several in the column) to apply a local magnetic field to deflect (rather than focus) the beam. To translate the beam we use deflector scan coils. To tilt the beam just before it reaches the specimen we use tilt scan coils situated between C2 and C3.

When we create a scanning beam for STEM imaging, the beam must always move parallel to the optic axis. Such scanning is accomplished by tilting the beam twice with two sets of scan coils, one above the other, to ensure that the beam crosses the optic axis at the front focal plane of C3. Then, wherever the beam enters the C3 lens field, it is bent to follow a path parallel to the optic axis. You can see how this is done if you look ahead to Figure 9.15. This rather complex adjustment is computer-controlled. Like many other procedures on a modern TEM, this adjustment is made automatically when you select STEM mode.

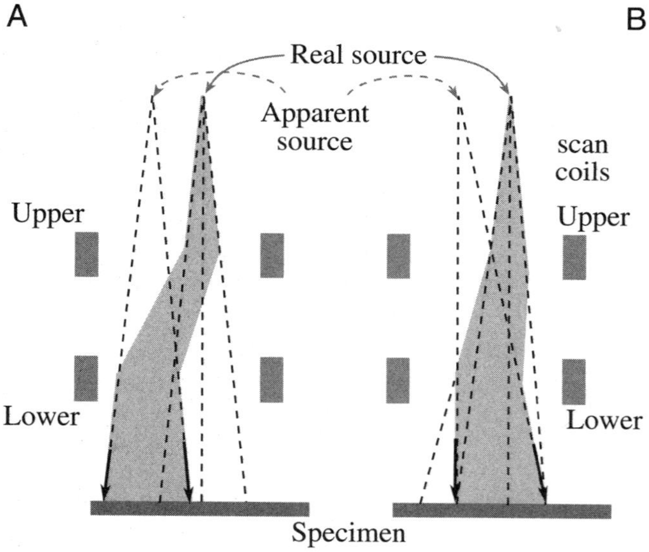

Figure 9.6. The use of pre-specimen scan coils for (A) traversing the beam and (B) tilting the beam. Traversing moves the beam to a different area of the specimen but it stays parallel to the optic axis. Conversely, tilting the beam illuminates the same area of the specimen, but from a different angle.

9.1.D. Alignment

If you correctly align the illumination system, the gun crossover is on the optic axis and the electrons can then follow a straight line through the lenses and apertures until they hit the specimen. Alignment used to be a tedious manual affair involving tilting and translating the gun and the condenser lenses and centering the apertures on axis. Now most of the components are machined accurately enough that minor electronic adjustment is all that is needed. Nevertheless, manual centering of the C2 aperture remains a most critical step in obtaining the best performance out of the TEM, particularly if you intend to operate in scanning mode for STEM imaging and microanalysis.

Instructions for alignment vary for different TEMs so we'll simply describe the principles. Even if you won't be doing the alignment you will want to check that it is correctly aligned; you can recognize if the wheels are not aligned on your car and you know it is important to balance them for best performance of the vehicle, even if you have to ask someone else to perform the task. If you want the best out of your machine, you'll want to be able to fine tune this alignment.

Gun Alignment

First, you have to undersaturate the filament so structure can be seen in the image, as shown back in Figure 5.5. If the gun is very badly misaligned, you may have to turn the condenser lenses off, before you use the gun traverses to center the filament image. Then use the gun tilts to make the source image symmetrical and repeat the whole procedure. If this alignment is very bad, then either there is a major problem with the gun or with the previous user; in either case expert help is required to correct the fault.

Alignment of the C2 Aperture

You must have the C2 aperture accurately centered on the optic axis of the TEM. If the aperture is misaligned the image of the beam on the screen moves off axis and distorts as you underfocus or overfocus C2, as shown in Figure 9.7. To align the aperture on axis, you need first to overfocus C2 so the image of the beam is spread and the outline of the C2 aperture is visible on the screen (make sure any other apertures in the imaging system are out of the column). Then use the external drives to center the aperture on the screen. Next, you must adjust C2 so the image of the beam is focused. Then, center the beam with the deflector controls. Now underfocus the C2 lens until you can again see the aperture and center it again with the external drives. You have to repeat this whole operation iteratively until the image of the beam expands and contracts around the center of the screen, as shown in Figure 9.8. Usually there's a control that will introduce an AC current into the lens coil, in effect "wobbling" the lens setting either side of focus. This saves you from manually underfocusing and overfocusing the lens.

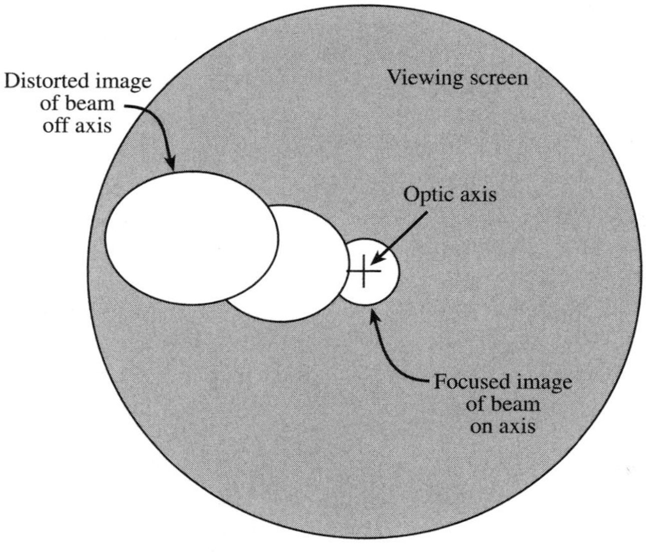

Figure 9.7. If the C2 aperture is misaligned, underfocusing or overfocusing the C2 lens causes the image of the beam to sweep off axis (i.e., across the viewing screen) and to become distorted.

9.1.E. Condenser Lens Defects

The illumination system lenses suffer from the standard lens defects, such as aberrations and astigmatism. These defects don't really limit the operation of the TEM in paral-

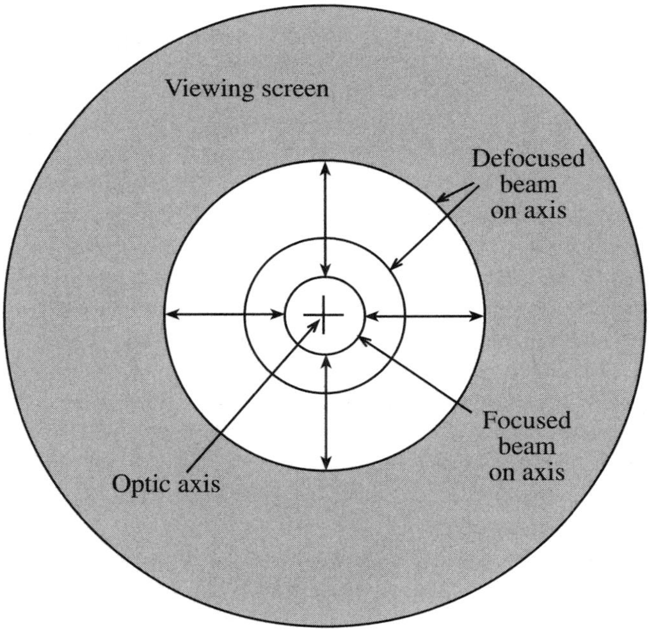

Figure 9.8. If the C2 aperture is aligned, the image of the beam remains circular and expands or contracts about the optic axis as the lens is underfocused or overfocused.

lel-beam mode, but they are crucial if you're intent on forming the finest probe possible for STEM and analytical work. Let's look at the role of each of the major defects.

Spherical Aberration

This defect plays no role in limiting parallel-beam formation. However, as we discussed in Chapter 5, in adjusting the illumination system to form the finest possible probe with the maximum available current, spherical aberration in the probe-forming lens (C3) controls the minimum possible probe size. In exactly the same manner as we control the image resolution (see Chapter 6), spherical aberration limits the probe dimensions to a minimum radius (equation 6.23) of $r_{min} \sim 0.91(C_s\lambda^3)^{1/4}$. This is why the C3 probe-forming lens has a short focal length (to minimize C_s). The final probe-limiting aperture in C2 needs to be carefully chosen to be the optimum value (equation 6.22) for the selected probe size $\alpha_{opt} = 0.77\ \lambda^{1/4}/C_s^{1/4}$. In practice, however, there are always more C1 settings than available C2 apertures, so it is not possible to choose the optimum aperture for each probe. This can cause problems if you need a specific probe size for a certain spatial resolution, as we discuss in Chapter 36.

Chromatic Aberration

Remember this aberration depends on the energy spread of the electrons. Since the electrons in the illumination system have such a small energy spread, you can regard them as monochromatic and there is no detectable degeneration of the probe dimensions.

Astigmatism

This is the most common defect in the TEM illumination system and arises either because the final C2 limiting aperture is misaligned or it is contaminated and charging up, thus deflecting the beam. Let's assume you've centered the C2 aperture as we just described and talk about correcting any residual astigmatism due to contaminated apertures.

You can discern astigmatism in the illumination system if you look at an image of the electron source on the screen; focus C2 so the beam is a minimum diameter and the image of the beam is circular, as you did when aligning the aperture. If you then wobble the C2 lens either side of the focal setting, the image of the beam expands and contracts about its minimum dimension. If there is astigmatism, the image is not circular, but distorts elliptically and rotates through 90° either side of focus, as shown in Figure 9.9. The condenser stigmators introduce a compensating field which you use to correct this distortion. You first overfocus the beam so you can see the effect of the astigmatism (i.e., the beam appears elliptical). Then adjust the stigmators so the image appears circular. Now underfocus the beam and repeat the correction. Repeat the whole over/underfocus procedure iteratively until the image of the beam remains circular as you expand and contract it on the screen with the C2 lens.

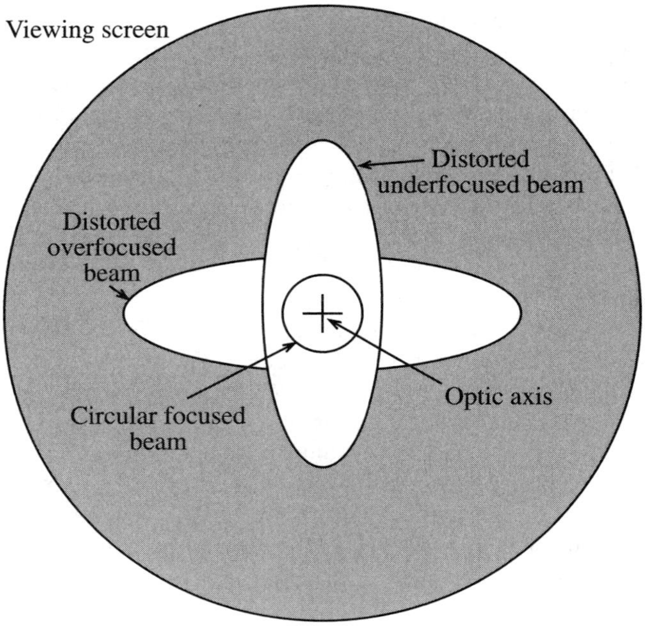

Figure 9.9. The effect of astigmatism in the illumination system is to distort the image of the beam elliptically as the C2 lens is underfocused or overfocused. Correction of this astigmatism results in an image that remains circular as the C2 lens is defocused.

If you can't make the image circular, you'll have to increase the range of strength of the stigmators. If you are on maximum strength, then you need to remove the source of the astigmatism by flame cleaning the condenser diaphragms, as we described in Chapter 6.

9.1.F. Calibration

We've already seen what it takes to calibrate the performance of the electron gun and optimize the brightness so that the maximum beam current goes into the minimum beam size. All that's left is to calibrate the condenser system. The major variables are the probe size for various C1 settings and the convergence angle for various C2 aperture sizes.

The C1 lens strength controls the probe size at the plane of the specimen. We've described in some detail how to measure the beam dimensions at the specimen back in Section 5.5.C. Figure 9.10 shows the variation of the calculated (not measured) probe size as a function of the C1 lens setting for a typical TEM. These calculations are approximate, since they define the probe width as the FWHM and

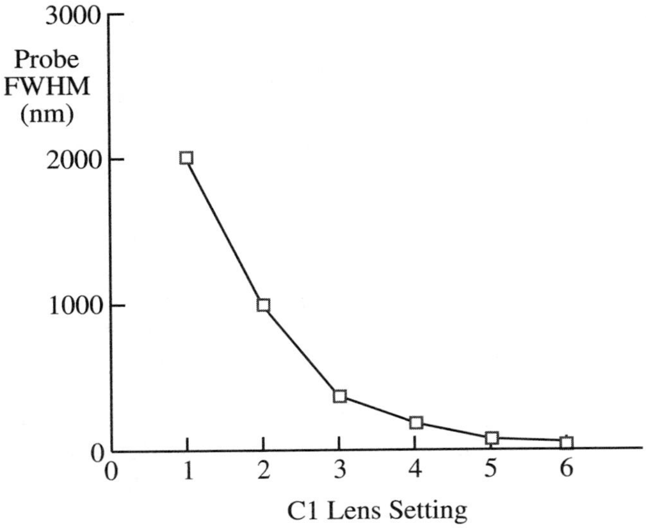

Figure 9.10. Calibration of the illumination system requires determining the variation of the probe size with C1 lens strength.

assume the C2 aperture correctly limits the Gaussian distribution. Despite these approximations you can clearly see the expected trend of decreasing probe size with increasing C1 strength.

The C2 aperture size governs the convergence semiangle α, as we also discussed in Chapter 5 when we were determining the gun brightness.

- We measure the total convergence angle 2α from a CBED pattern (Figure 5.8).
- We increase 2α by increasing the C2 aperture size (Figure 9.11).

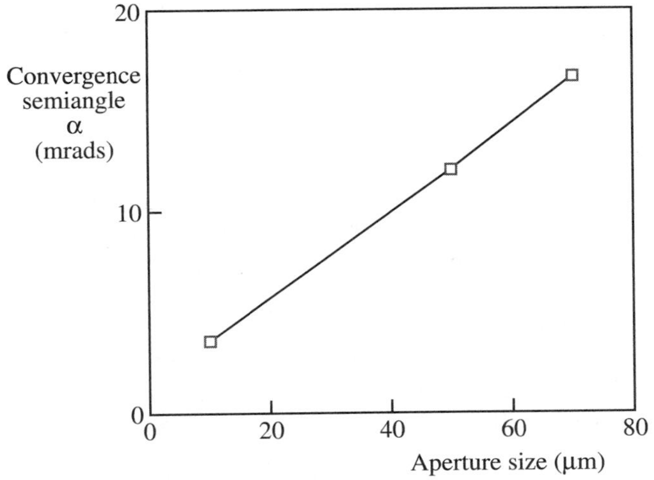

Figure 9.11. Variation of the beam-convergence semiangle, α, with the C2 aperture dimensions.

Different TEMs will have different responses. On some instruments the C2 aperture is virtual (so you have an effective aperture size), which makes it rather difficult to measure 2α. (See Goldstein *et al.* 1992 for a detailed description of this problem, which is common in SEMs.) Furthermore, if the C2 lens is excited, it can also change α and then you have to calibrate α as a function of both the aperture size and the C2 lens setting, which is an extremely tedious exercise.

9.2. THE OBJECTIVE LENS AND STAGE

This combination is the heart of the TEM. We use the stage to clamp the specimen holder in the correct position so the objective lens can form images and diffraction patterns in a reproducible manner. As we discussed in Chapter 8, there are two different types of holder, top-entry and side-entry, and these determine the geometry of the polepiece and the flexibility with which you have to make adjustments. Our discussion will emphasize the side-entry holder since this is becoming the standard, but top-entry holders require the same adjustment of the *z*-control or specimen height.

> We need to fix the height of the specimen on the optic axis. This will allow us to work at the same objective lens current and thus at a fixed objective lens magnification.

As a practical consideration, you would like to be able to tilt the sample without changing its height on the optic axis. Otherwise, you would be continuously using the *z*-control when you tilt the sample. Clearly, this means that you should ensure that the region of the specimen you want to work on is located close to the tilt axis of the specimen rod.

The central requirement is the need to define a reference plane so that our calibrations will be reproducible. The reference plane (see Chapter 6) for a side-entry holder is the *eucentric plane*. This plane is normal to the optic axis and contains the axis of the specimen holder rod; clearly there could be many such planes. What is special about the eucentric plane is that when the specimen is located at this plane and the image is in focus, the objective lens current is an optimum value. The position of this plane within the objective lens is known as the eucentric height. If you put your specimen in the eucentric plane, then a point on the optic axis does not move laterally when you tilt it around the holder axis. Of course, if you tilt your specimen normal to the holder axis, or rotate it off axis, then the point you're examining almost invariably moves out of the eucentric plane.

The first thing you must always do when inserting your specimen into the TEM is to ensure that it is in the eucentric plane. To do this, you tilt the specimen and adjust the height of the specimen holder until the image of the specimen remains stationary when you tilt the sample through ±30° either side of zero.

> With computer control and auto-focusing techniques becoming common, this operation can be automated. As a result we now see completely eucentric stages in which your specimen doesn't move off the optic axis and remains in focus no matter around what axis it is tilted or rotated. If you don't have a computer-controlled stage, be cautious.

The eucentric plane should also be coincident with the plane that is symmetrically positioned with respect to the upper and lower objective polepiece fields. This means that the eucentric plane coincides with the plane at which the electron beam is imaged, in both TEM and STEM modes. If the symmetric plane and the eucentric plane are not coincident, then the images and diffraction patterns will appear at different magnifications and different focus settings in TEM and STEM. Obviously this requirement has no meaning in a DSTEM where there is no TEM mode.

Ensuring coincidence of the eucentric and symmetric planes is usually carried out by the manufacturer. You can check it by comparing the focus of a diffraction pattern or an image in TEM and STEM modes. You should not have to refocus the image or diffraction pattern with the objective lens when you change from one mode to the other.

9.3. FORMING DIFFRACTION PATTERNS AND IMAGES: THE TEM IMAGING SYSTEM

You know that the objective lens takes the electrons emerging from the exit surface of the specimen, disperses them to create a diffraction pattern (DP) in the back focal plane, and recombines them to form an image in the image plane (see Figure 6.4). We can use this ray diagram to introduce the basic operations for forming static-beam images and diffraction patterns in the TEM. We'll then describe how to do the same thing with a scanning beam in STEM mode.

In this discussion we will skip many of the details and concentrate on the role of the instrument. In Chapter 11 we will discuss the details of the diffraction process and then expand these ideas in Chapters 16 through 21. We'll then discuss the images formed in the TEM in Chapters 22 through 31. The first operation that you need to master when using the TEM is viewing the diffraction pattern. In all the subsequent imaging, we'll use this pattern to select electrons that have suffered particular angles of scatter to form our images.

- To see the diffraction pattern you have to adjust the imaging system lenses so that the back focal plane of the objective lens acts as the object plane for the intermediate lens. Then the diffraction pattern is projected onto the viewing screen, as shown in Figure 9.12A.
- If you want to look at an image instead, you readjust the intermediate lens so that its object plane is the image plane of the objective lens. Then an image is projected onto the viewing screen, as shown in Figure 9.12B.

Let's look now at the details of these two fundamental operations from the point of view of the instrument. In subsequent chapters we will discuss how to understand the images and why we form them in the ways we do.

9.3.A. Selected-Area Diffraction

As you can see from Figure 9.12A, the diffraction pattern contains electrons from the whole area of the specimen that we illuminate with the beam. Such a pattern is not very useful because the specimen will often be buckled. Furthermore, the direct beam is often so intense that it will damage the viewing screen. So we perform a basic TEM operation both to select a specific area of the specimen to contribute to the diffraction pattern and to reduce the intensity of the pattern falling on the screen. If you look at Figure 9.12A, there are two ways we could reduce the illuminated area of the specimen contributing to the diffraction pattern.

- We could make the beam smaller.
- We could somehow insert an aperture above the specimen which would only permit electrons that pass through it to hit the specimen.

The first option involves using C2 and/or C3 to converge the beam at the specimen. We use this approach to form CBED patterns, which we'll discuss in great detail in Chapters 20 and 21. Converging the beam destroys any coherence, and spots in the pattern are not sharply defined but spread into disks. If we wish to obtain a diffraction pattern with a parallel beam of electrons, the standard way is to use a selecting aperture. This operation is called se-

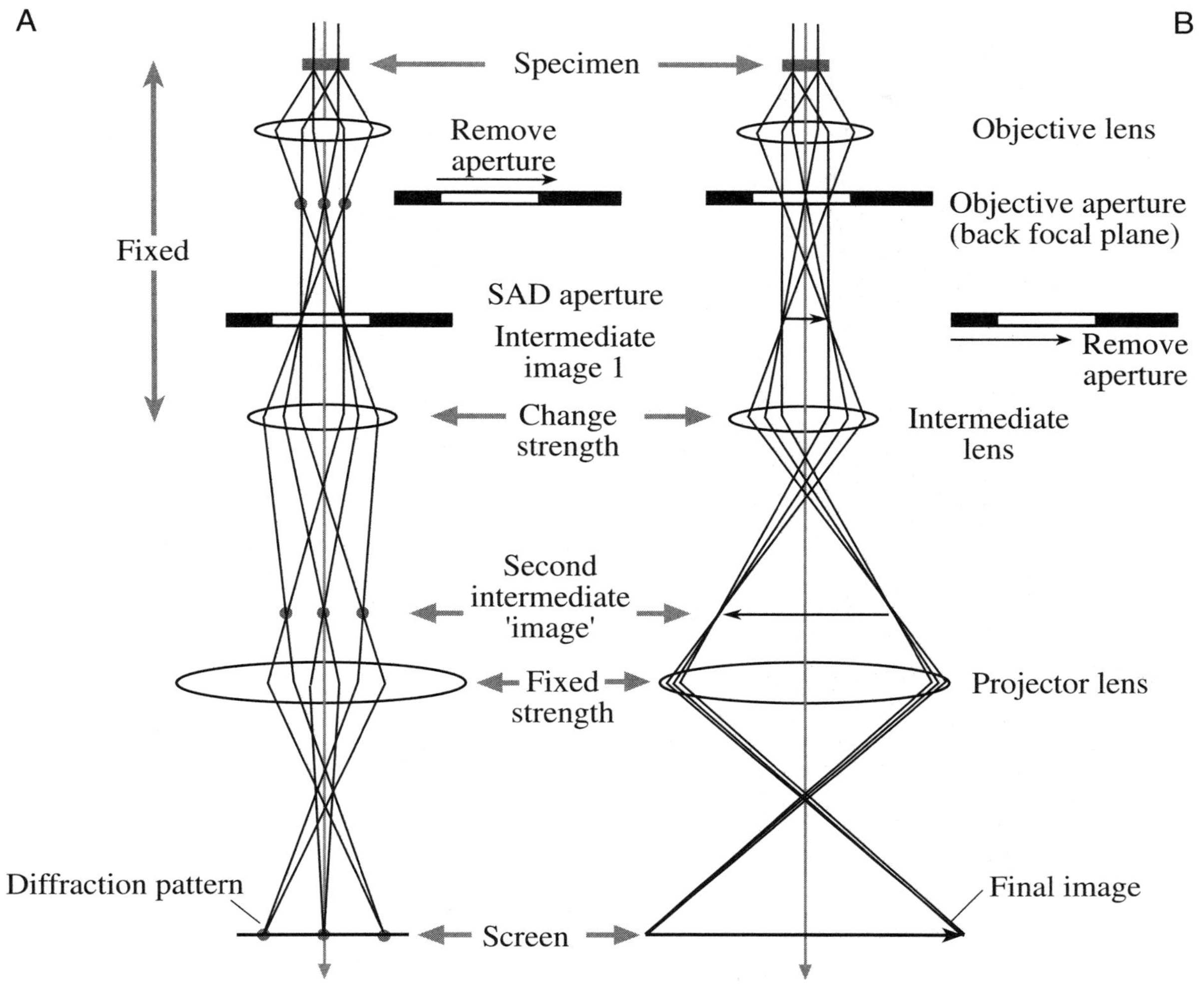

Figure 9.12. The two basic operations of the TEM imaging system involve (A) projecting the diffraction pattern on the viewing screen and (B) projecting the image onto the screen. In each case the intermediate lens selects either the back focal plane or the image plane of the objective lens as its object.

lected-area diffraction, or SAD, and was invented by LePoole (1947). Now, we can't insert an aperture at the specimen plane, because the specimen is already there! If we insert an aperture in a plane conjugate with the specimen, i.e., in one of the image planes, then it creates a virtual aperture at the plane of the specimen. This is exactly what we do.

The conjugate plane that we choose is the image plane of the objective lens, as shown in Figure 9.13. We insert the *SAD aperture* into the image plane of the objective lens and center the aperture on the optic axis in the middle of the viewing screen. You can see the image of this aperture on the viewing screen. It must be focused by adjusting the intermediate lens so it is conjugate with (i.e., exactly in the plane of) the image of the specimen that we focused with the objective lens. Then any electron that hits the specimen outside the area defined by the virtual aperture will hit the real diaphragm when it travels on to the image plane. It will thus be excluded from contributing to the diffraction pattern that is projected onto the viewing screen. In practice, we can't make apertures smaller than about 10 μm, and the demagnification back to the plane of the specimen is only about 25×, which gives a minimum selected area of ~0.4 μm—which isn't as small as we'd like. We'll discuss in Chapter 11 whether or not smaller values would be useful.

The SAD pattern is often displayed on the viewing screen at a fixed magnification.

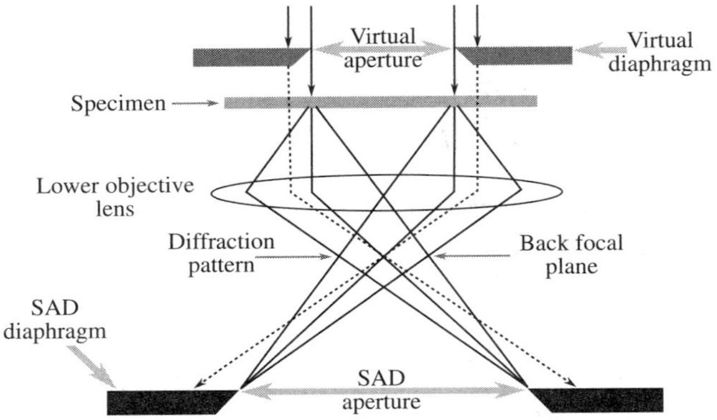

Figure 9.13. Ray diagram showing SAD pattern formation: the insertion of an aperture in the image plane results in the creation of a virtual aperture in the plane of the specimen. Only electrons falling inside the dimensions of the virtual aperture at the specimen will be allowed through into the imaging system. All other electrons will hit the SAD diaphragm.

> By analogy with the hand-held camera we define a distance called the "camera length" (*L*).

This distance corresponds to the distance of the "film" from the diffraction pattern. We choose the value of *L* such that the spacings in the diffraction pattern are easily discernible on the screen and on the photographic plate. This magnification can be changed by adjusting the intermediate lenses. We'll describe how we calibrate this magnification later.

> It is a basic principle of TEM operation that when you want to look at the diffraction pattern (i.e., the *back focal plane* of the objective lens), you put an *SAD aperture* into the *image plane* of the objective lens.

You can see this aperture if you want to change it or center it, by projecting the image plane onto the viewing screen, which we'll now discuss.

Beware: In most TEM books, SAD is the only standard diffraction technique. As a result, some microscopists use only SAD to obtain diffraction information. However, you should know that CBED, which we discuss in Chapters 20 and 21, can provide complementary diffraction information and must also be used by all TEM operators in the materials field. There are still certain times when you'll need to form an SAD pattern.

- When you need to select a spot from which to form a BF or DF image (see next section).
- When diffraction spots are very close to one another and would overlap in CBED patterns (see examples in Chapters 23 and 24).
- When you are looking for fine structure in the diffraction pattern, such as streaks (see Chapter 17).

On all other occasions, when the diffraction maxima provide the most important information in the pattern, then we strongly recommend that you use CBED.

9.3.B. Bright-Field and Dark-Field Imaging

When the SAD pattern is projected onto the viewing screen, we can use the pattern to perform the two most basic imaging operations in the TEM. No matter what kind of specimen you're are looking at, the SAD pattern will contain a bright central spot which contains the direct electrons and some scattered electrons (as shown in Figures 2.11a–c). When we form images in the TEM, we either form an image using the central spot, or we use some or all of the scattered electrons. The way we choose which electrons form the image is to insert an aperture into the back focal plane of the objective lens, thus blocking out most of the diffraction pattern except that which is visible through the aperture. We use the external drives to move the aperture so that either the direct electrons or some scattered electrons go through it. If the direct beam is selected as shown in Figure 9.14A, we call the resultant image a bright-field (BF) image, and if we select scattered electrons of any form, we call it a dark-field (DF) image, as shown in Figure 9.14B. Typical magnification ranges will be 25,000× – 100,000×.

The BF and DF images can be viewed at any magnification simply by adjusting the intermediate lenses of the microscope. It is necessary to calibrate the actual magnification and also to be able to relate directions in the image at any magnification to directions in the diffraction pattern at a fixed camera length. These are the two basic calibrations required for any TEM.

> It is another principle of TEM operation that if you want to view an image (i.e., the *image plane* of the objective lens) you insert an aperture into the *back focal plane* of the objective lens. This is called the *objective aperture* and is most important in the TEM, since its size controls the collection angle (β) and hence determines the effect of all the aberrations and resolution of the most important lens in the instrument.

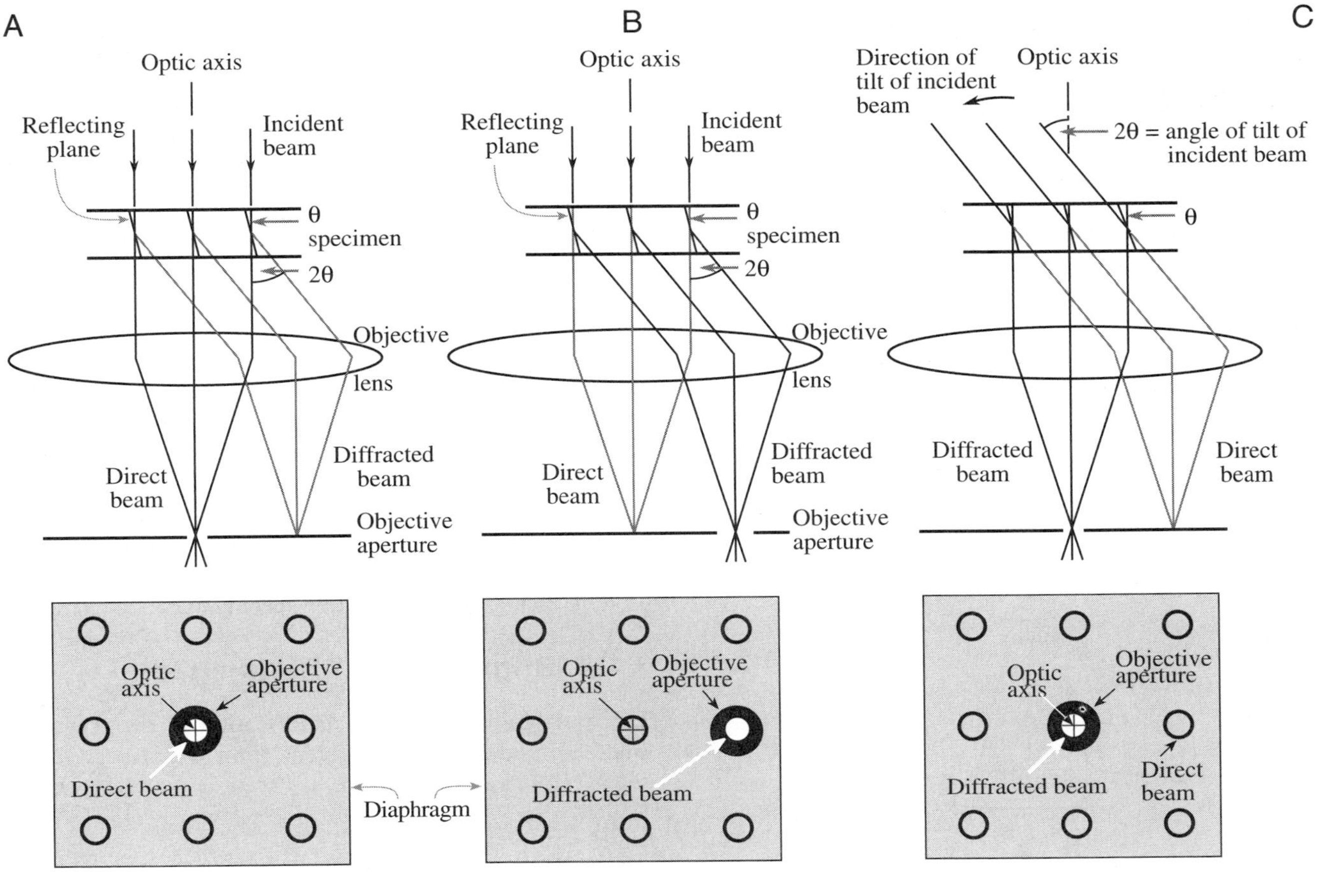

Figure 9.14. Ray diagrams showing how the objective lens/aperture are used in combination to produce (A) a BF image formed from the direct beam, (B) a displaced-aperture DF image formed with a specific off-axis scattered beam, and (C) a CDF image where the incident beam is tilted so that the scattered beam remains on axis. The area selected by the objective aperture, as seen on the viewing screen, is shown below each ray diagram.

The insertion and removal of the SAD and objective apertures can be confusing to the beginner and often the wrong aperture is inserted, or not removed when it should be. You have to practice obtaining SAD patterns and BF/DF images to get used to what aperture should be inserted and when. Both apertures are inserted below the objective lens. The objective aperture goes into the back focal plane, so it is closer to the lens (i.e., higher up the column) than the SAD aperture which is in the image plane. Remember that if you're looking at a diffraction pattern, the (lower) SAD aperture should be inserted and the (upper) objective aperture removed. If you want to look at an image, the objective aperture should be inserted and the SAD aperture removed.

9.3.C. Centered DF Operation

If you look at Figure 9.14B, the electrons that are selected by the aperture travel off the optic axis, since we displace the aperture to select the scattered electrons. These off-axis electrons suffer aberrations and astigmatism and the DF image is difficult to focus, since it will move on the screen as you adjust the objective lens strength. To avoid this you have to adjust the beam tilt potentiometers above the objective lens so that the incident beam hits the specimen at an angle equal and opposite to the scattering angle. In this way the scattered electrons will now travel down the optic axis, as shown in Figure 9.14C. This operation is called centered darkfield (CDF) imaging and is the way to do DF imaging in the TEM, if you want to record the best, focused image. However, there are situations where you will want to form a displaced-aperture DF image. You do this by physically moving the aperture rather than by tilting the incident beam.

We'll return to BF, CDF, and SAD operations when we discuss specific contrast mechanisms that occur in TEM images in Chapter 22.

9.4. FORMING DIFFRACTION PATTERNS AND IMAGES: THE STEM IMAGING SYSTEM

If you want to use a fine probe to form STEM images, then the objective lens optics are a little more complex than in TEM. The key feature to remember is that the scanning beam must not change direction as the beam is scanned.

> The beam has to scan parallel to the optic axis at all times so that it mimics the parallel beam in a TEM even though it's scanning.

As we show in Figure 9.15, the way we do this is to use two pairs of scan coils to pivot the beam about the front focal plane of the upper objective (C3) polepiece. The C3 lens then ensures that all electrons emerging from the pivot point are brought parallel to the optic axis and an image of the C1 lens crossover is formed in the specimen plane. Now, if the objective lens is symmetrical, then a stationary diffraction pattern is formed in the back focal plane (so called because this pattern does not move even though the beam is scanning, since it is conjugate with the front focal plane, as shown in Figure 9.16). If we stop the beam from scanning, then we have a CBED pattern in the back focal plane and we can project that onto the TEM screen if we wish. Now let's discuss how to form STEM images.

One potentially very big advantage of forming images this way is that we don't use lenses, as in an SEM. So defects in the *imaging lenses* do not affect your image resolution, which is controlled by the beam only. Hence chromatic aberration, which can limit TEM images, is absent in STEM images, which is advantageous if you're dealing with a thick specimen. However, there are drawbacks also, as we'll discuss below, and STEM images aren't widely used, particularly for crystalline specimens.

9.4.A. Bright-Field STEM Images

The basic principle of image formation in the scanning mode is fundamentally different from that for a static-beam TEM image. As you've just seen, in the TEM we se-

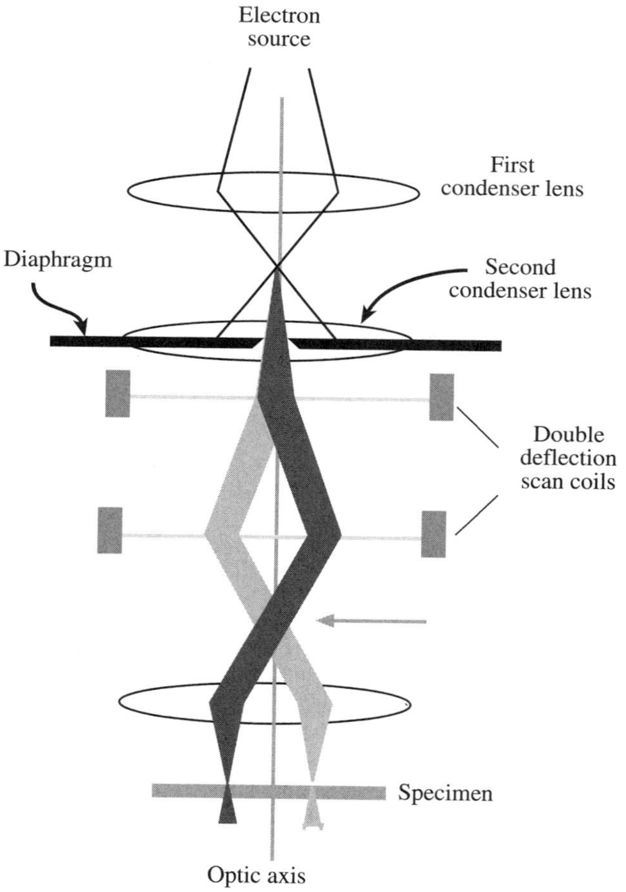

Figure 9.15. Scanning the convergent probe for STEM image formation using two pairs of scan coils between the C2 lens (usually switched off) and the upper objective polepiece. The probe remains parallel to the optic axis as it scans.

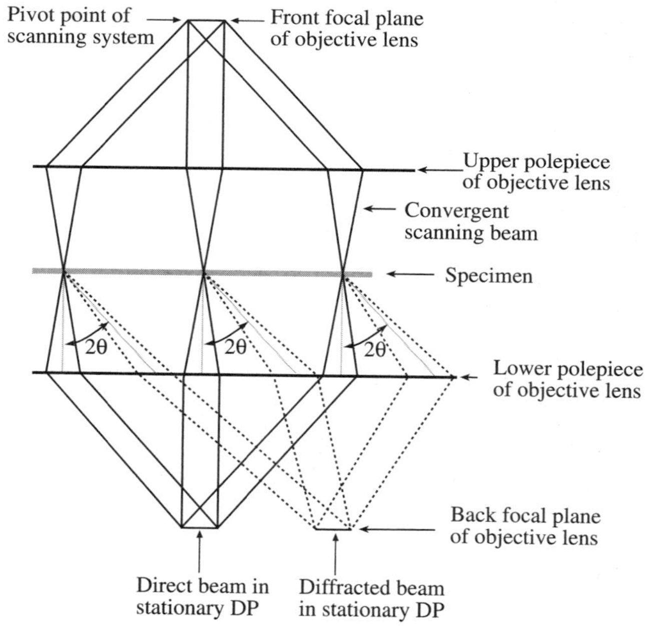

Figure 9.16. The creation of a stationary (convergent-beam) diffraction pattern in the back focal plane of the objective lens is a necessary prerequisite for STEM imaging. Note that electrons scattered through 2θ at different points in the specimen are focused at the same point in the focal plane.

lect a portion of the electrons emerging from an area of the specimen and project that distribution onto a screen. The principle of scanning image formation is shown in Figure 9.17. Simply stated, the beam is scanned on the specimen by adjusting the scan coils; these same coils are used to scan the CRT synchronously. The electron detector acts as the interface between the electrons coming from the specimen and the image viewed on the CRT. Since it takes up to 2048 scan lines to build up an image on the CRT, the whole process of creating a STEM image is much slower than TEM imaging: it's serial recording instead of parallel recording.

> The STEM signal generated at any point on the specimen is detected, amplified, and a proportional signal is displayed at an equivalent point on the CRT. The image builds up over several seconds or even minutes.

This process is exactly the same principle as used in any scanned-probe microscope, such as an SEM or an STM (scanning tunneling microscope). Remember that to form a TEM BF image, we inserted an aperture into the plane of the TEM diffraction pattern and only allowed the direct electrons through it into the imaging system. In STEM mode we use an electron detector, in exactly the same way as we use the aperture: we only allow the electrons that we want to contribute to the image to hit the detector. So we put a BF detector (either a semiconductor or scintillator detector) into the direct beam in the diffraction pattern in Figure 9.18. Thus, only direct electrons hit that detector from wherever the beam is scanning on the specimen. This variable signal travels from the detector via an amplification system to modulate the signal on a CRT, thus creating a BF image as also shown in Figure 9.18.

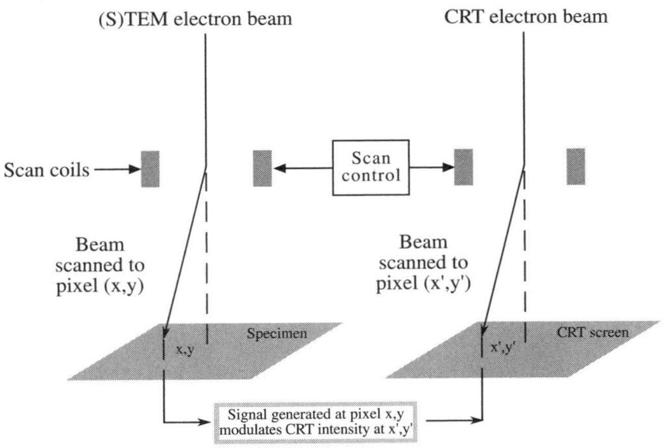

Figure 9.17. The principle of forming a scanning image, showing how the same scan coils in the microscope control the beam-scan on the specimen and the beam-scan on the CRT. Thus no lenses are required to form the image.

> The variable signal which emerges from the BF detector depends on the intensity in the direct beam from that point on the specimen.

Now, in a TEM we can't physically put the detector in the back focal plane of the objective lens to form a STEM image, because it would interfere with the objective aperture. Therefore, we usually insert the detector into a conjugate plane to the diffraction pattern, below the projector lens. So when you form a STEM image in a TEM, you operate the TEM in diffraction mode and insert a detector into the viewing chamber of the TEM, either above or below the screen. The stationary diffraction pattern falls on the detector and the signal goes to the CRT. In a DSTEM, there may not be any imaging system lenses, in which case the detector is positioned immediately after the objective lens. Much of what we've just said is automatically done when you "hit the STEM button." The message is the same: understand what is happening and why.

9.4.B. Dark-Field STEM Images

The approach is analogous to that of TEM. We form a DF image by selecting any of the scattered electrons, rather than the direct electrons. Remember, in a TEM we tilt the incident beam so the scattered electrons that we want to form the image travel down the optic axis and are selected by the objective aperture. In a STEM, we do things rather differently.

> If we want a specific beam of scattered electrons to fall on the BF detector, we can simply shift the stationary diffraction pattern so that the scattered beam is on the optic axis.

It's simple to do this with the diffraction pattern centering controls, or you could also displace the C2 aperture. The former is to be preferred, since doing the latter misaligns the illumination system.

9.4.C. Annular DF Images

Rather than using the BF detector for DF imaging, we usually use an annular detector, which surrounds the BF detector, and then all the scattered electrons fall onto that detector. We call this annular dark-field (ADF) imaging and it has certain advantages, depending on the contrast mechanism operating in the specimen, as we'll see in Chapter 23.

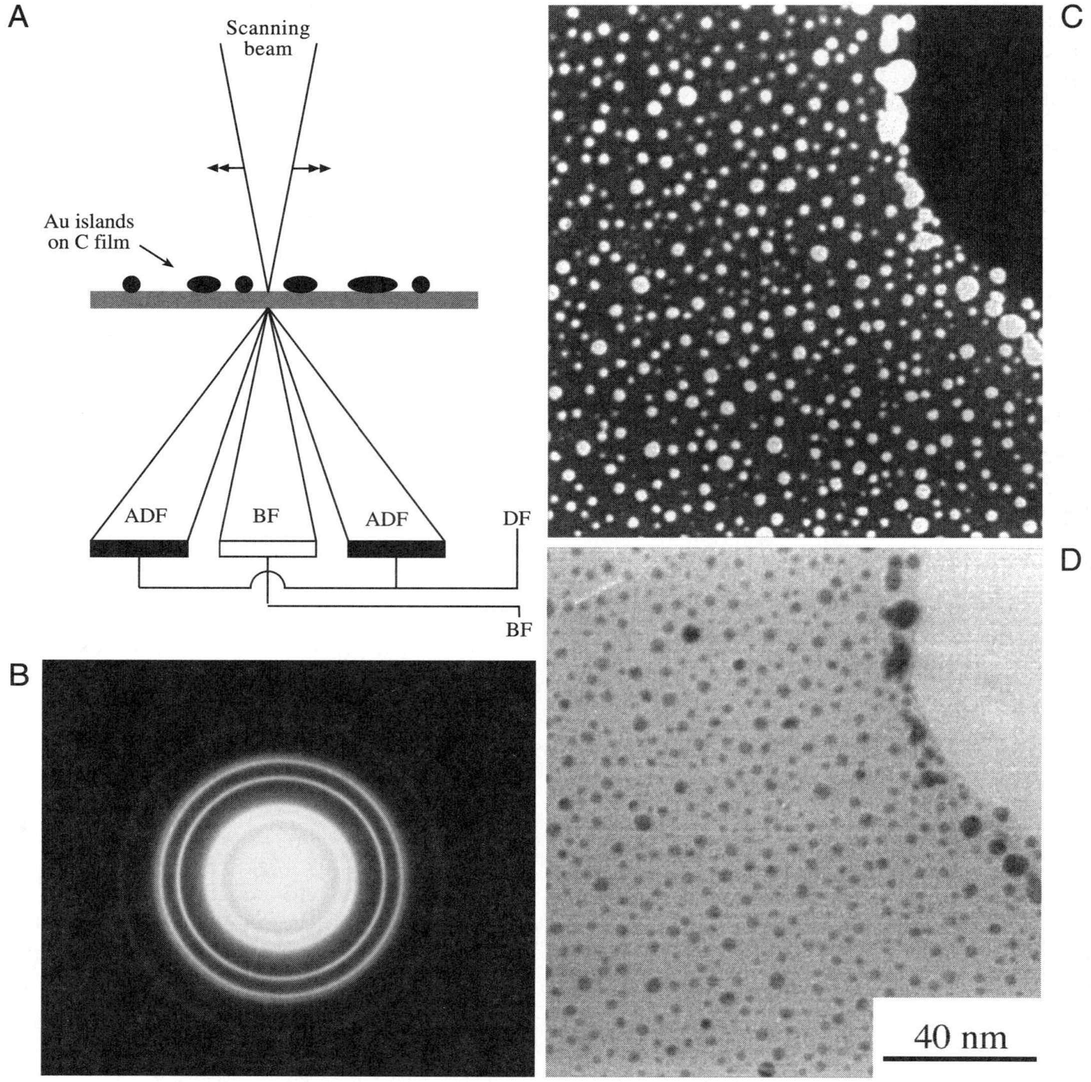

Figure 9.18. STEM image formation: A BF detector is placed in a conjugate plane to the back focal plane to intercept the direct beam (A) and a concentric annular DF detector intercepts the diffracted electrons (B). The signals from either detector are amplified and modulate the STEM CRT. The specimen (Au islands on a C film) gives complementary ADF (C) and BF (D) images.

As is shown in Figure 9.18, the ADF detector is centered on the optic axis and has a hole in the middle, within which the BF detector sits. The resultant ADF image in this simple example is complementary to the BF image.

Now, of course, you can make the detector any size or shape you wish. For example, you can split the annulus into two halves or four quadrants and electrically isolate each part of the detector. Then you can form images from electrons that fall on different parts of the detector. It's impossible to do this in a TEM, because the objective aperture that does the selecting is a hole and can't be cut up. We'll talk more about these kinds of detectors when we discuss specific contrast mechanisms in TEM and STEM images in Chapter 23.

9.4.D. Magnification in STEM

Any of the STEM images that we have just described appear on the CRT screen at a magnification that is controlled by the scan dimensions on the specimen, *not* the lenses of the TEM. This is a fundamental difference between scanning and static image formation. Scanning images are *not* magnified by lenses (and are thus not affected by aberrations in the imaging lenses as we stated at the start of this section). If the scanned area on the specimen is 1 cm × 1 cm, and the resultant image is displayed on a CRT with an area 10 cm × 10 cm (which is the standard size of the *record* CRT, though rarely the size of the *viewing* CRT screen) then the magnification is 10×. If the scan dimension is reduced to 1 mm, the magnification is 100×, and so on, up to magnifications in excess of 10^6×, which are common in dedicated STEMs. As with the TEM, we have to calibrate the STEM magnification and the camera length of the diffraction pattern we use to create the images.

9.5. ALIGNMENT AND STIGMATION

9.5.A. Lens Rotation Centers

You only need to perform two alignments to ensure that the imaging system is operating correctly. By far the most important is the alignment of the objective lens center of rotation, and the second is the alignment of the diffraction pattern on the optic axis. To get the best out of your TEM, you must master these two fundamental alignments.

Basically, the idea of the objective lens rotation alignment is to ensure that the objective lens field is centered around the optic axis, so that the direct electrons emerging from the specimen see a symmetric field as they pass through the lens. If the field is off center, then the electrons will move off axis and your image will rotate about a position off axis as you change the objective lens (focus), as shown schematically in Figure 9.19.

To center the objective rotation, you start at a relatively low magnification (say 10,000×), select an obvious reference point in the image, and observe the way the point rotates as you wobble the objective lens over and under focus. Then use the beam tilts to move the center of rotation to the middle of the screen and repeat the process at higher magnifications. Above ~100,000× the wobbler may introduce too large a rotation, so you may have to defocus the objective lens manually. The actual steps to do this are instrument-dependent, so consult the manufacturer's handbook. This process is also called "current centering."

In some instruments you can also perform "voltage centering" in which a varying voltage is applied to the gun and the objective lens is aligned to ensure that the electrons remain on axis through the lens as their energy varies. Not all instruments are capable of this alignment.

The diffraction center is aligned by adjusting the projector lens until the central spot in the diffraction pattern is on axis. If you change the diffraction pattern magnification (the camera length) the pattern will move off axis, which can easily be compensated for in a computer-controlled column.

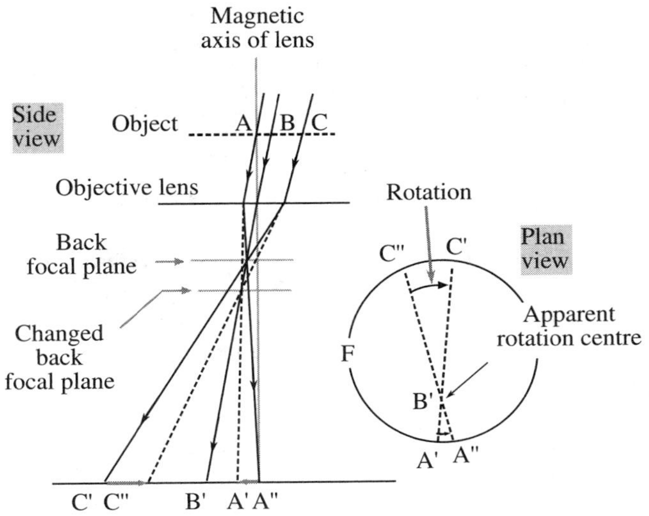

Figure 9.19. When the objective lens center of rotation is misaligned, the image appears to rotate about a point away from the center of the viewing screen when the lens is wobbled about focus.

Centering the diffraction pattern is useful in STEM image formation, since you use it to center the diffraction pattern on the STEM detector such that the direct beam hits the BF detector and the scattered beams hit the ADF detector. Apart from this simple operation the STEM imaging system needs no lens alignment.

9.5.B. Correction of Astigmatism in the Imaging Lenses

After you've centered the image and diffraction pattern, the main cause of problems in the imaging system is objective and intermediate lens astigmatism.

Objective lens astigmatism occurs mainly if the objective aperture is misaligned, so you must carefully center the aperture on the optic axis, symmetrically around the electron beam used to form the BF or DF image. Despite careful centering, however, residual contamination may cause astigmatism and then you have to use the objective stigmators to introduce a compensatory field.

You'll find that the effects of objective astigmatism are harder to see than condenser astigmatism, which is easily visible on the screen. Often you can only see objective astigmatism at the highest magnifications, where it manifests itself as a streaking in the image that rotates through 90° as you alternately underfocus and overfocus the objective lens. So again, you have to wobble the objective lens, but if the magnification is too high then manual wobbling is required.

> If you then overfocus C2 to ensure a parallel beam, and also overfocus the objective lens, then a Fresnel fringe will be visible at the thin edge of the specimen.

Either look for a small hole in the specimen or look at the edge of the specimen. Ideally, you might use a holey carbon film to correct residual astigmatism *before* you put in your specimen, especially while learning this procedure. In practice, you have to check your astigmatism throughout your TEM session so you use the same approach on a thin curved edge of your specimen.

As shown in Figure 9.20A, when you underfocus the objective lens, there is a bright fringe round the edge of the hole. If this fringe is uniform around the hole, then there is no astigmatism. If the fringe varies in intensity as in Figure 9.20D, then the focus of the lens is changing around the hole because of astigmatism. So you then have to adjust the objective stigmators to make the fringe uniform. The same operation must be repeated at overfocus, when there is a dark fringe around the edge of the hole (Figure 9.20B). At exact focus, there is no fringe and the image contrast is minimized (Figure 9.20C).

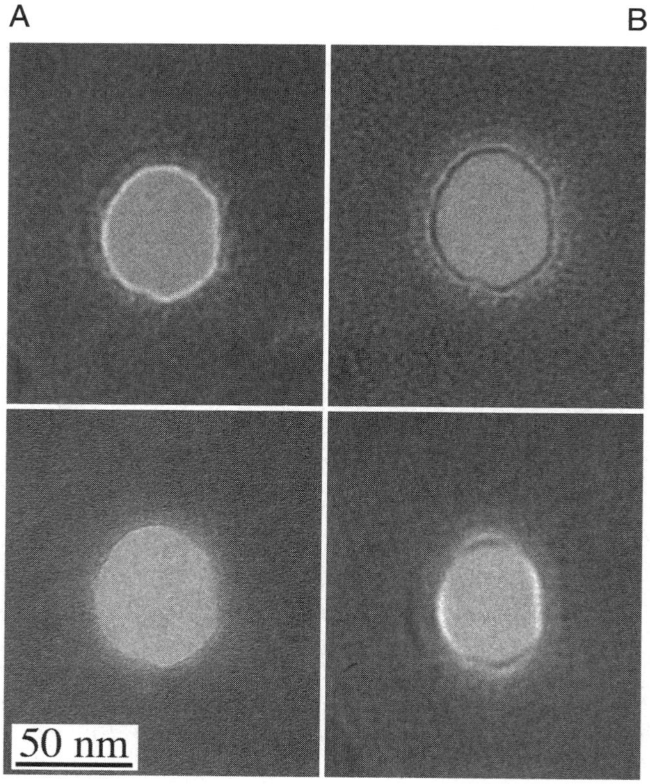

Figure 9.20. The image of a hole in an amorphous carbon film illuminated with a parallel beam showing that (A) with the beam underfocused, a bright Fresnel fringe is visible; (B) with the beam overfocused a dark fringe is visible; (C) at exact focus there is no fringe; and (D) residual astigmatism distorts the fringe.

This method of correcting the astigmatism is reasonable at magnifications up to several hundred thousand times. For high-resolution imaging at magnifications of >300,000×, we actually use the streaking in the image to correct for astigmatism. We'll talk about this when we discuss HRTEM in Chapter 28.

Intermediate lens astigmatism is of secondary importance and only affects the DP. Because the DP is at zero magnification in the objective lens, the intermediate lenses are responsible for magnifying it. So if there is residual astigmatism in these lenses, then the DP will show orthogonal distortions as you take it through focus. This effect is small and can only be seen in the binoculars as you focus the DP with the diffraction focus (intermediate lens) control. Make sure that the incident beam is strongly underfocused to give the sharpest spots. As with objective astigmatism in the image, simply adjust the intermediate stigmators to compensate for any spot distortion at underfocus, and overfocus until the spots expand and contract uniformly in all directions through focus. You should be aware that not all instruments have the requisite intermediate stigmators to carry out this correction.

9.6. CALIBRATION OF THE IMAGING SYSTEM

Your TEM should be calibrated when it is first installed and then periodically throughout its life, especially if you wish to carry out accurate measurements from images or diffraction patterns. If the instrument is modified substantially, then it must be recalibrated. In all cases you must specify a set of standard conditions under which the calibrations are carried out (e.g., objective lens current and other lens settings, eucentric height, etc.).

> Since you usually will not be the first user, you should take the time to check the existing calibration. Don't assume it is correct.

9.6.A. Magnification Calibration

We use standard specimens to calibrate the magnification. The most common specimens we use are thin carbon film replicas of optical diffraction gratings of known spacing, such as shown in Figure 9.21A. The typical linear density

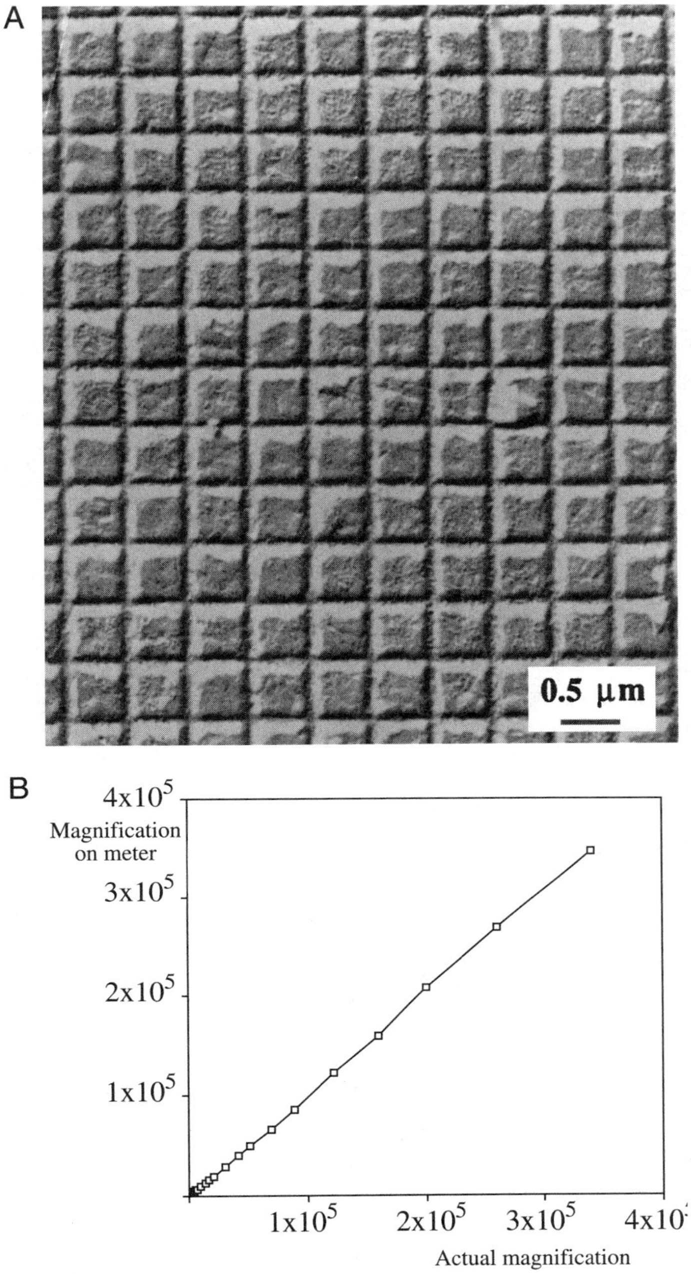

Figure 9.21. (A) An image of a diffraction grating replica in which the actual spacing of the grating is known. (B) The TEM magnification can thus be calibrated, relating specific magnification settings to be assigned specific magnifications.

of lines in the replica is 2160 lines per mm (giving a line spacing of 0.463 μm), which enables calibration up to magnifications of about 200,000×. Above this magnification, individual grating spacings are wider than the film. So we then use small latex spheres (50–100 nm diameter) although they are susceptible to beam damage and shrinkage under electron bombardment. At the highest magnifications, the images of known crystal spacings, such as the 0002 spacing in the graphite structure (0.344 nm), can be used. What we are doing is just using a known periodicity in the crystal; careful consideration of the objective lens defocus and specimen thickness is required before the phase-contrast lattice image can be directly interpreted, as we discuss in detail later in the text.

Magnification calibration is so sensitive to so many variables that some users deposit a standard material on the material they are studying so that the calibration will be done under exactly the same conditions and will appear on the same negative.

Basically, you set the TEM to its standard conditions with the specimen at the eucentric height and focus the image. Next, record images of the diffraction grating at all magnification settings and calculate the magnification experimentally from the image. Figure 9.21B shows the magnification calibration for a Philips CM30 TEM, using both a diffraction grating and latex spheres.

You have to calibrate the magnification because the TEM imaging system does not give stable and reproducible lens strengths. The lens strengths will change with ambient temperature, with the efficiency of the cooling system of the lenses, and with lens hysteresis. Therefore, if you want to make accurate measurements from TEM images, you must carry out the magnification calibration at the time you make the measurements. In particular, you have to minimize the lens hysteresis by always approaching image focus consistently from overfocus or underfocus and/or reversing the lens polarity several times before finally coming to focus. Also, you must remember that there may be distortions in the image, particularly at low magnification (<5000×). You can find a full description of all the details concerned with TEM calibration in Edington (1976).

Because of the magnification error in the TEM, it is not the best instrument for absolute measurement of parti-

Table 9.1. Magnification Calibration for a Philips EM400T in STEM Mode at 120kV

Digital readout	Calculated magnification
3,200	3,420
6,400	6,850
12,500	12,960
25,000	27,000
50,000	54,000
100,000	108,000

cle sizes, etc. However, relative measurement is easily done with reasonable accuracy (±5%), so long as you note the precautions we've just described. Without a calibration the digital readout is probably no better than ±10% accurate, and so it is unwise to state magnifications to better than ±10%. You should be suspicious of any micrographs that you see in the literature with a magnification that is more precise than this (e.g., 52,700×). It may indicate that the microscopist does not understand the instrument's limitations and the work should be interpreted with due caution.

> Remember that the electromagnetic lenses have hysteresis and the area of the specimen you are working on must be at exactly the right "height" in the column.

You can use an identical procedure to calibrate the STEM CRT image magnification. This is equally important despite the fact that the digital STEM image magnification is, in principle, easily calculated from the scan coil strengths. The image magnification differs from the digital readout because of variations in the objective lens. Table 9.1 shows the difference between a typical digital readout of the STEM magnification and the experimentally determined magnification using a diffraction grating replica.

9.6.B. Camera-Length Calibration

We describe the magnification of the diffraction pattern by the camera length (L), a term that arises from X-ray projection diffraction cameras which operate without lenses (because focusing X-rays is very difficult). In these cameras magnification is increased by moving the recording film further away from the specimen. This principle can be applied in the TEM, as shown in Figure 9.22. This figure represents the imaging system, but without the lenses drawn in.

> If we increase the magnification of the lenses between the specimen and the viewing screen, we increase the effective distance L between the specimen and the screen.

The camera length in the TEM is thus a calculated value rather than a physical distance. If electrons are scattered through an angle 2θ at the specimen (as in a typical diffraction event), then the separation of the direct and diffracted beams as measured on the screen (R) is determined by L since

$$\frac{R}{L} = \tan 2\theta \sim 2\theta \qquad [9.1]$$

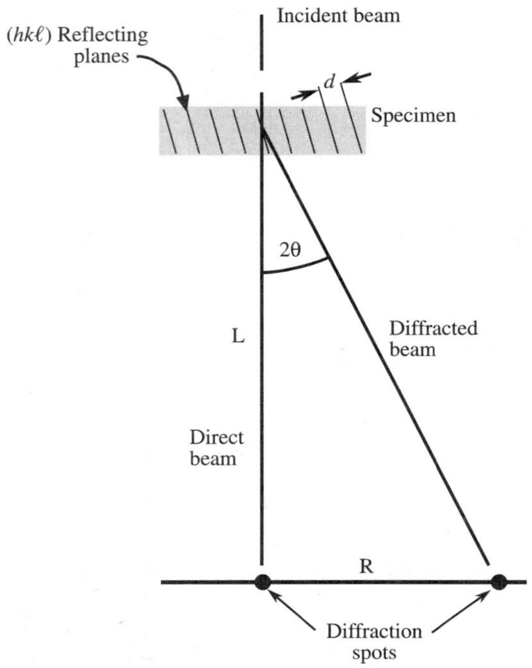

Figure 9.22. The relationship between the spacing R of diffraction maxima and the camera length, L. Increased magnification corresponds to effectively increasing L, although in practice this is accomplished with lenses.

From the Bragg equation we know that $\lambda/d = 2 \sin\theta \sim 2\theta$, and so we can write

$$Rd = \lambda L \qquad [9.2]$$

Thus to calibrate the magnification of the diffraction pattern we need to record patterns from a specimen with known crystal spacing (d), such as a thin film of a polycrystalline Au or Al. This gives a ring pattern (see Figure 2.5). We know the lattice parameter of the specimen, we can measure the ring radius R on the photographic film for any plane that is diffracting (see Chapter 11 to find out exactly how we do this), and since we know λ we can determine L. A typical TEM camera length calibration is shown in Table 9.2. The STEM camera length calibration may be different than the TEM if the objective lens setting is not exactly the same in TEM and STEM modes, and this depends on the vintage and make of your instrument. So you should check with the manufacturer before taking the time to perform the calibration.

> As a general rule, you should *always* do the calibrations yourself, and not rely on any factory calibrations, because the conditions you use in your laboratory may differ from those of the manufacturer.

Table 9.2. Comparison of Experimentally Measured Camera Length (and Camera Constant) with the Digital Readout for a Philips EM400T Operating at 120 kV (λ = 0.0335Å)

Camera length Setting	Digital readout (mm)	Measured camera length, L (mm)	Camera constant λL (mm Å)
1	150	270	9.04
2	210	283	9.47
3	290	365	12.22
4	400	482	16.14
5	575	546	18.28
6	800	779	26.08
7	1150	1084	36.29
8	1600	1530	51.22
9	2300	2180	72.99
10	3200	3411	114.20

9.6.C. Rotation of the Image Relative to the Diffraction Pattern

Anyone studying crystalline materials must determine the angle between directions in the image and directions in the diffraction pattern. At a fixed camera length, the diffraction pattern always appears on the screen in a fixed orientation. But if you record images at different magnifications, the images will rotate by an angle Φ with respect to the fixed diffraction pattern. (In some TEMs this rotation has been removed by the addition of a compensating projector lens, and in this case there is always a fixed rotation, ideally 0°, between common directions.)

To determine this rotation, we use a specimen of α-MoO_3, because it forms thin asymmetric crystals with a long edge known to be parallel to the 001 direction in the crystal. You must take care to ensure that, as usual, the image is focused with the specimen at the eucentric plane. Then insert the SAD aperture and ensure that it is focused using the intermediate lenses to coincide with the image plane. Finally, switch to diffraction mode with the beam underfocused and adjust the diffraction focus to give sharp diffraction maxima. Then, take a double exposure of the diffraction pattern and the image as shown in Figure 9.23A. Repeat the whole exercise for different magnifications and plot out the variation of the angle Φ as shown in Figure 9.23B. You can do the same if necessary for different values of L, which introduce a systematic change in Φ. It is recommended that you carry out all your SAD work at a standard value of L; 500–1000 mm is usually optimum.

A further complicating factor is that, as the image magnification is increased, the TEM lens control logic may switch off, or switch on, one of the imaging system lenses. When this happens, a 180° inversion is introduced into the image. You can see this happen if you watch the image carefully as you change the magnification. This inversion has to be included in the rotation calibration otherwise a 180° error will be made in the assignment of directions in the image. One way to see if the image has a 180° inversion is to look at the diffraction pattern and defocus it slightly so the BF image in the direct beam can be seen directly at very low magnification. The 180° inversion is immediately obvious, as shown in Figure 9.24.

9.6.D. Analysis of TEM Images and Diffraction Patterns

If you don't use a double exposure when comparing images and diffraction patterns (or indeed when comparing directions in any two films), you need a fixed reference line. This line must be independent of slight variations that may arise depending on the film size, how you loaded it, etc. The best reference line is the edge of the plate numbering system that is superimposed on each film.

> Whenever you're comparing images and diffraction patterns, it is essential to compare the photographic negatives with the *emulsion side up*. This is contrary to usual photographic practice, but it's necessary to preserve the relationship between manipulations of your specimen and what you see happening on the screen. If you don't do this, it is easy to introduce a 180° error into the relationships between images and diffraction patterns.

9.7. OTHER CALIBRATIONS

The accelerating voltage: The selected voltage may differ from the absolute voltage by detectable amounts. There are several ways to determine the actual voltage: First you can

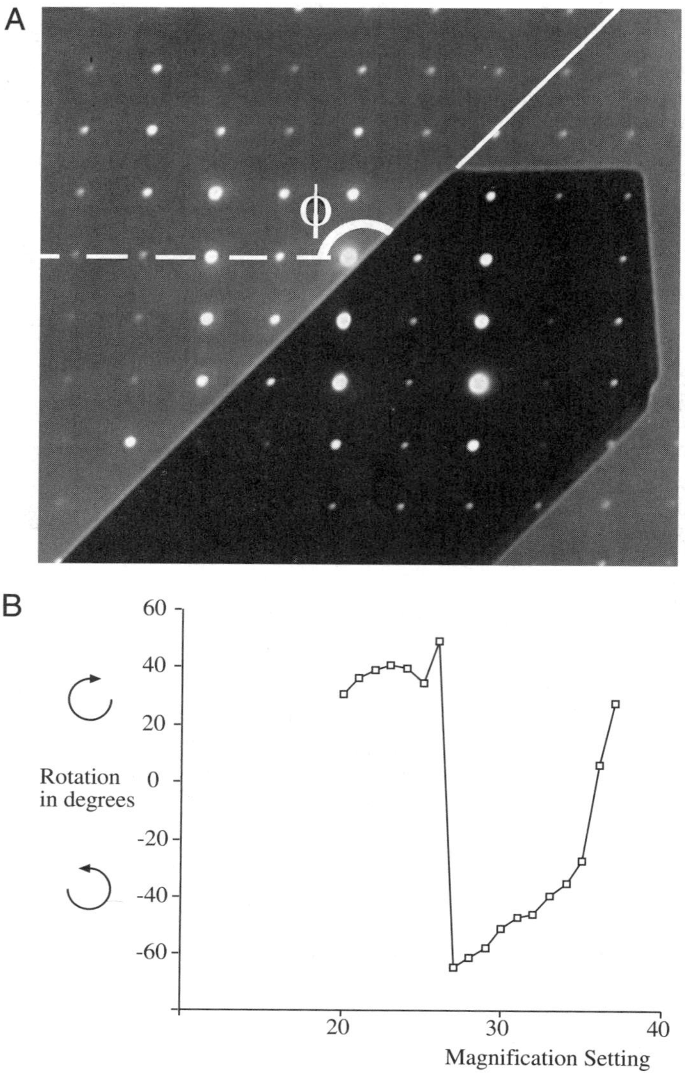

Figure 9.23. (A) A double exposure showing the superposition of an image of a MoO_3 crystal on a diffraction pattern from the same crystal, defining the rotation angle Φ. (B) The rotation calibration gives the angle Φ between equivalent directions in the image and the diffraction pattern as the magnification is varied. The calibration assumes a constant camera length.

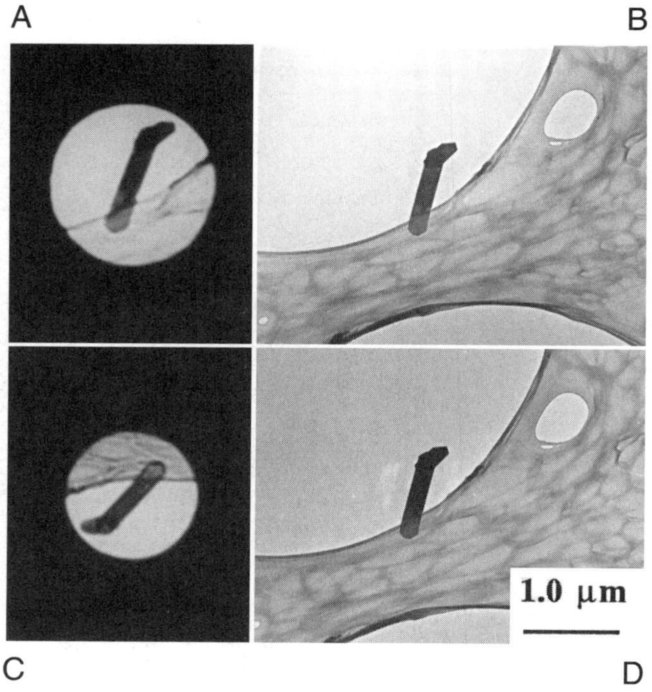

Figure 9.24. Defocused direct beam in a diffraction pattern from MoO_3 compared with a BF image, showing how to determine if a 180° inversion exists or not. If the image in the spot is rotated with respect to the image on the screen, as in (C) and (D), then the 180° inversion is required. In (A) and (B), no rotation occurs between the DP and BF image.

measure the electron wavelength λ by measuring the angle ϕ between two Kikuchi line pairs (see Chapter 19) which intersect at a distance R from the direct beam

$$\tan \phi = \frac{R}{L} = \frac{\lambda}{d} \qquad [9.3]$$

Alternatively, you can match simulations of higher-order Laue-zone lines to experimental lines in CBED patterns (see Chapter 21) and determine which λ gives the best match. Finally, if you can get your X-ray computer system to display the X-ray spectrum (Chapter 34) out to E_0, the beam energy, the bremsstrahlung intensity vanishes to zero at the exact beam energy (this is called the Duane-Hunt limit).

The specimen tilt axis and the sense of tilt: In a side-entry stage, the principal tilt axis is parallel to the specimen holder rod. Since the image is often rotated relative to the specimen, how can you locate the rotation axis? Move the specimen. From this movement, you can determine the sense of tilt for a specimen of known geometry.

> If you gently push on the end of your side-entry specimen holder, the image moves parallel to the principal tilt axis.

If you are looking at the diffraction pattern, defocus the pattern so you can see the BF image in the zero spot, as in Figure 9.25, then carry out the same exercise. If you are using a top-entry holder, you will need to calibrate this tilt using a known specimen geometry.

Focal increments of the objective lens: If you're going to do high-resolution phase-contrast imaging, then

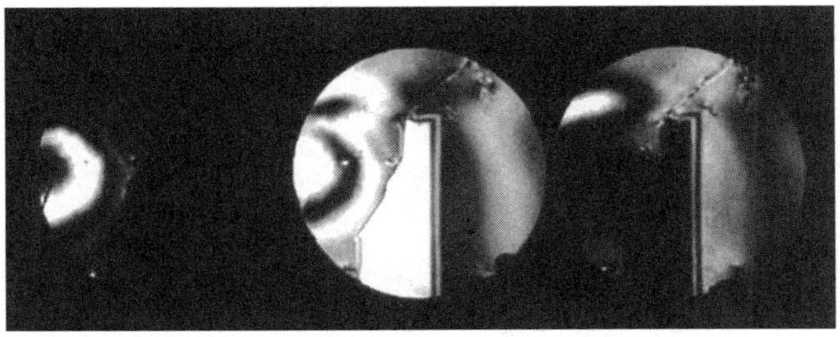

Figure 9.25. Defocused multiple DF image showing how it is possible to determine simultaneously the direction of features in the image (e.g., the vertical twin boundary) and directions in the DP (e.g., the horizontal vector between the diffraction disks). If the specimen holder is moved in the direction of the principal tilt axis, the image will move and identify the relationship between that tilt axis and the DP.

you need to know the value of each defocus step of the objective lens. There is a simple method for determining this step value. Superimpose a focused image and an image defocused by a known number of objective lens focal increments (Δf). The two images will be separated by a distance Δx which is related to Δf by

$$\Delta f = \frac{\Delta x}{2Mm\theta} \qquad [9.4]$$

where M is the magnification, m is the number of focal increments, and θ is the Bragg angle for the reflection used to form the image. If you use some typical values, you'll find that it is difficult to be very accurate with this method. We'll return to this topic in Chapter 28.

CHAPTER SUMMARY

We've now shown you how a TEM is put together. While the manufacturer does a pretty good job, there are still some essential steps for you, the operator, to carry out. You must understand how to align the illumination system so the beam is on axis. You can then create a parallel beam for TEM and a convergent one for STEM. The C2 aperture is a crucial part of the whole illumination system and the most easily misaligned. Astigmatism is not too much of a problem if the instrument is kept clean. The objective lens/stage combination controls all the useful information that is created as the beam is scattered by your specimen. *Always* start a microscope session by fixing the eucentric height, and before you do any worthwhile imaging, align the objective center of rotation and minimize the astigmatism at high magnification. Diffraction and STEM operation require a centered diffraction pattern.

If you want to make any quantitative measurements from your images and diffraction patterns (and you really ought to do this if you have any aspirations to be a real microscopist), then calibration cannot be avoided. Your images and diffraction patterns are relatively useless unless you know their magnification (camera length) and the angular relationship between the two. So take the time to do this early in your studies. In doing so you will not only ensure that you produce quality data, but you will also learn an enormous amount about how these complex instruments work.

REFERENCES

General References

Chapman, S.K. (1980) *Understanding and Optimizing Electron Microscope Performance 1. Transmission Microscopy,* Science Reviews Ltd., London.

Chapman, S.K. (1986) *Maintaining and Monitoring the TEM,* Royal Microscopical Society Handbook, No. 8, Oxford University Press, New York.

Chescoe, D. and Goodhew, P.J. (1990) *The Operation of Transmission and Scanning Microscopes,* Royal Microscopical Society Handbook, No. 20, Oxford University Press, New York.

Specific References

Edington, J.W. (1976) *Practical Electron Microscopy in Materials Science,* Van Nostrand Reinhold, New York.

Goldstein, J.I., Newbury, D.E., Echlin, P., Joy, D.C., Romig, A.D. Jr., Lyman, C.E., Fiori, C.E., and Lifshin, E. (1992) *Scanning Electron Microscopy and X-ray Microanalysis,* Plenum Press, New York.

LePoole, J.B. (1947) *Philips Tech. Rundsch.* **9**, 93.

Specimen Preparation

10

CHAPTER PREVIEW

There are many ways to prepare specimens for the TEM. The method you choose will depend on both the type of material and the information you need to obtain. One important point to bear in mind is that your technique must not affect what you see or measure, or if it does then you must know how. Specimen preparation artifacts may be interesting but they are not usually what you want to study.

Specimen preparation is a very broad subject; there are books devoted to this topic alone. The intention here is to summarize the techniques, suggest routes that you might follow, and above all to emphasize that there are many ways to produce a TEM specimen; the one you choose will depend on the information you need, time constraints, availability of equipment, your skill, and the material. So we'll concentrate on the "principles of cooking," but won't try to list all the possible "recipes."

The TEM specimen, when you've made it, must be electron transparent and representative of the material you want to study. In most cases (not all) you would like your specimen to be uniformly thin, stable under the electron beam and in the laboratory environment, conducting, and nonmagnetic (we'll discuss some exceptions as we proceed). Few specimens approach the ideal and usually you have to compromise. In general we can divide specimens into two groups: self-supporting specimens and specimens resting on a support grid or thin washer; the grid is usually Cu but could be Au, Ni, Be, etc. Before discussing these two groups we will briefly review the most important part of specimen preparation, namely safety. You may damage the microscope later, but this is the stage where you could do much worse to yourself.

It is often assumed that preparation of the TEM specimen will take several hours. Actually, this time could be as short as five minutes or as long as two days, even for the same material. For example, as you'll see, if you want to examine a piece of $YBa_2Cu_3O_{6+x}$, the high-temperature superconductor, you could crush the sample in a mortar and pestle using a nonaqueous solvent, catch the small particles on a carbon film, and put the specimen in the TEM; time required, about ten minutes. Alternatively, you might cut the sample into thin slices using a diamond saw, cut 3-mm-diameter disks from the slice, thin the disk on a grinding wheel, dimple the thinned disk, then ion mill to electron transparency at liquid-nitrogen temperatures, carefully warm the specimen to room temperature in a dry environment and put it in the TEM; time required, one or two days. Which method you choose would depend on what you want to learn about your material.

Specimen Preparation

10

10.1. SAFETY

Either the specimen itself or the best method for preparing it for viewing in the TEM may require extreme care. Even specimens which are safe and relatively inert in bulk form may be hazardous in powder form. Four favorite (because they work so well) liquids for polishing solutions are hydrogen cyanide, hydrofluoric acid, nitric acid, and perchloric acid. These liquids may be poisonous, corrosive (HF penetrates the body to dissolve the bone), or explosive (perchloric acid and nitric acid when mixed with certain organic solvents). It is clearly essential that you check with your laboratory manager, the reference texts, and the appropriate material safety data sheets (MSDS) before you begin specimen preparation. You might also save a lot of time.

In spite of these restrictions, you may still need/want to use these acids and acid/solvent mixtures. The ion thinner may not be available or you may not be able to accept the damage which ions produce. In this event there are five brief points that you should bear in mind:

- Be sure that you can safely dispose of the waste product *before* you start.
- Be sure you have the "antidote" at hand.
- Never work alone in the specimen preparation laboratory. Always wear safety glasses when preparing specimens and/or full protective clothing, including face masks and gloves, if so advised by the safety manual.
- Only make up enough of the solution for the one polishing session. Never use a mouth pipette for measuring any component of the solution. Dispose of the solution after use.
- Always work in a fume hood when using chemicals. Check that the extraction rate of the hood is sufficient for the chemical used.

Since these four acids can be so dangerous, we'll mention them specifically, but remember—always seek advice before chemically preparing specimens.

Cyanide solutions: If possible avoid this solution even though you may see it in the textbooks. The only metal where it really excels is gold and you can thin this by very careful ion milling.

Perchloric acid in ethanol or methanol: If you have to use this "universal polish" you should be aware that many laboratories require that you use a special dedicated hood which can be completely washed down, since crystallized perchloric acid is explosive. The phase diagram in Figure 10.1 for the perchloric–acetic (acid)–water system makes the message clear. If you have to use perchloric–acetic acid mixtures, or indeed when using any perchloric-containing mixtures, keep the density below 1.48. If you are very careful, if you *always* add the acid to the solvent, and you make sure that the liquid *never* becomes warm, then perchloric acid solutions can be used to produce excellent TEM specimens of Al, stainless steel, and many other metals and alloys.

Nitric acid: In combination with ethanol, this acid can produce explosive mixtures, especially if left for long periods of time and exposed to sunlight. It is preferable to use methanol rather than ethanol, but in either case, keep the mixture cool and dispose of it properly.

HF: This acid is widely used in the semiconductor industry and in "frosting" light bulbs; the reason in both cases is that it dissolves SiO_2 leaving no residue. Careful use of dilute solutions can produce specimens which have large thin areas. If you use HF, completely cover any ex-

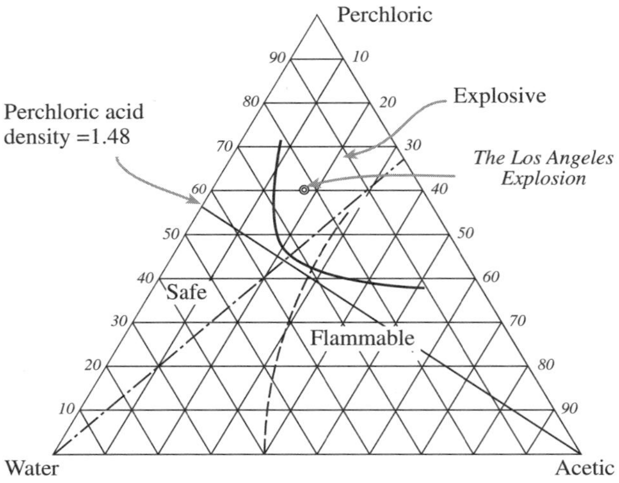

Figure 10.1. Perchloric–acetic–water phase diagram showing the hazardous regions and the recommended density line for safe use of all perchloric solutions. Always operate to the left of this line. (After Medard *et al.* 1949)

posed skin; HF rapidly penetrates the flesh and dissolves bone.

10.2. SELF-SUPPORTING DISK OR USE A GRID?

The type of TEM specimen you prepare depends on what you are looking for, so you need to think about the experiment that you are going to do *before* you start thinning. For example, is mechanical damage to be avoided at all costs, or can it be tolerated so long as chemical changes don't occur—or vice versa? Is the specimen at all susceptible to heat or radiation? Depending on the answers to these questions, some of the following methods will be inappropriate.

A self-supporting specimen is one where the whole specimen consists of one material. Other specimens are supported on a grid or on a Cu washer with a single slot. Several grids are shown in Figure 10.2. Usually the specimen or grid will be 3 mm in diameter.

Both approaches have advantages and disadvantages. Both offer you a convenient way of handling the thin specimen, since either the edge of the self-supporting disk or the grid will be thick enough to pick up with tweezers. If possible, never touch your specimen when it is thin. We recommend vacuum tweezers, but you'll need to practice using them; you can quite easily vibrate the specimen and break the thin area. Mechanical stability is always crucial. For example, single crystals of GaAs or NiO break very easily, so it is usually an advantage to have your specimen mounted on a grid since then you "handle" the grid. However, if you are performing X-ray analysis on a specimen the grid may contribute to the signal, because the X-rays can also arise from the grid. Thus you see a Cu peak where no Cu is present in the specimen. We'll talk later about how to minimize this artifact. Of course, the self-supporting specimen essentially has the same problem—it's just not as obvious! In fact, the preferred geometry for such analysis is usually the one where the specimen is thinnest.

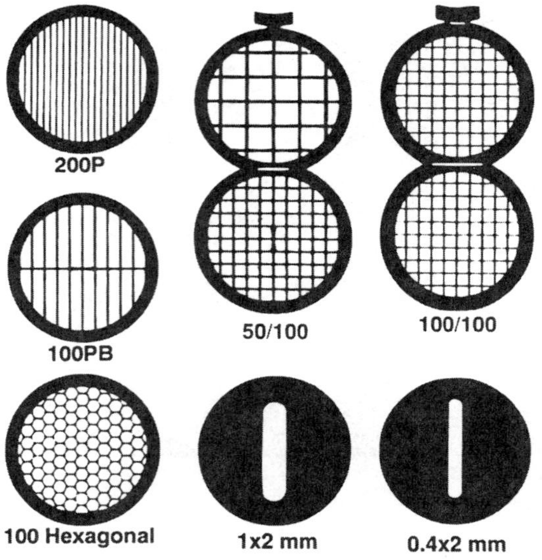

Figure 10.2. A variety of specimen support grids of different mesh size and shape. At the top right is the oyster grid, useful for sandwiching small slivers of thin material.

Why 3-mm disks? The disk diameter is usually a nominal 3.05 mm. We thus refer to the specimen as a 3-mm disk. Occasionally you will encounter a microscope which uses a 2.3-mm disk. The smaller diameter was used in earlier microscopes and has two important advantages, which are not fully exploited by modern machines. Ideally, the region of the specimen which you want to study will be located at the center of your disk no matter how large the disk is. As we saw in Chapter 9, the reason is that as you tilt the specimen in the microscope, the region of interest will then stay at the same position (height) above the objective lens and on the optic axis. Since for a self-supporting disk the rim of the specimen must be relatively thick and the total area of the material you'll study is small and confined to the center of the disk, you can make more 2.3-mm specimens from a given volume of material. This may be very important if the specimen is particularly special (expensive, rare) or if specimens break easily. A specimen which is 5 mm x 5 mm will give one 3-mm disk or four 2.3-mm disks. The second advantage of such specimens relates to tilting; the specimen holder can be manufactured to allow a greater tilt angle. Don't forget that if you only need one axis of tilt you may find the bulk holder useful. Then you can use a specimen which may be up to 1 cm long and 3 mm wide.

10.3. PREPARING A SELF-SUPPORTING DISK FOR FINAL THINNING

Final thinning involves three parts:

- Initial thinning to make a slice of material between 100 μm and 200 μm thick.
- Cut the 3-mm disk from the slice.
- Prethin the central region from one or both faces of the disk to a few micrometers.

The method you use will depend on what you want to study and the physical characteristics of the material (whether it is soft or hard, ductile or brittle, delicate or robust, single-phase or a composite, etc.).

10.3.A. Creation of a Thin Slice from the Bulk Sample

The materials you may need to thin can vary enormously. Clearly, we have to treat ductile and brittle materials differently.

(a) *Ductile materials such as metals.* Usually you don't want to introduce mechanical damage. For example, you want to study the defect structure or the density of defects in processed materials. The ideal method is to use a chemical wire/string saw, a wafering saw (not diamond—the soft metal will dull the blade), or spark erosion (electro-discharge machining) to get a thin slice < 200 μm. (A string saw works by passing the string through an acid or solvent and then across the sample until the string "cuts" through the sample; for example, you can use dilute acid to cut copper.) You could also roll the material to very thin sheet, then anneal it to remove the defects introduced by rolling.

(b) *Brittle materials such as ceramics.* Here there are two cases: (i) where you must not introduce mechanical damage, (ii) where you don't mind introducing mechanical damage or the material won't damage. You have several options depending on the material. Some materials (Si, GaAs, NaCl, MgO) can be cleaved with a razor blade; these are materials with a well-defined cleavage plane and it is possible to carry out repeated cleavage to electron transparency (see later). In other cases the ultramicrotome (see below) allows you to cut very thin slices for immediate examination in the TEM. If you don't want to cleave the specimen or you want to prepare a specimen parallel to a plane that doesn't cleave, you will need to use a diamond wafering saw. There are special techniques for some materials: you can, for example, use water as the solvent on a string saw to cut rocksalt. One of the main limitations with sawing is that the process destroys some of your sample.

10.3.B. Cutting the Disk

The same constraints hold as for cutting slices: if the material is reasonably ductile and mechanical damage is not crucial, then the disks can be cut using a mechanical punch (Figure 10.3). A well-designed punch can cut disks with only minimal damage around the perimeter, but the shock can induce shear transformations in some materials. For more brittle materials the three principal methods are spark erosion, ultrasonic drilling, and a grinding drill. In each case the cutting tool is a hollow tube with an inner diameter of 3 mm. Again, you want the wall of the tube to be thin to minimize the amount of material which is wasted. Spark erosion is used for conducting samples and introduces the least amount of mechanical damage. The choice between an ultrasonic drill (vibrating in H_2O) and a grinding (or slurry) drill is often a matter of personal preference or availability. Both remove material mechanically and are

Figure 10.3. A mechanical punch for stamping disks from thin sheets of ductile materials. A sheet sample is placed in the punch as indicated and the handle on the right is pushed down, ejecting a 3-mm-diameter disk suitable for thinning

widely used for ceramics and semiconductors. The drill may leave small particles in the specimen, and *all* mechanical thinning methods leave some surface damage. As a rule of thumb, abrasives produce damage to 3× their grit size. So a 1-μm abrasive will cause damage to 3 μm below the surface of each side of the specimen. Hence the final disk must be thicker than 2× the damage depth or else mechanical damage will be visible in the final specimen.

Note that there are variations for all these techniques; e.g., for Si, GaAs, and some other materials you can glue the sample to a support, coat it with a protective layer, and cut circles through the film—then chemically etch the desired region. You need to experiment, but the method introduces no damage.

10.3.C. Prethinning the Disk

The aim of this process is to thin the center of the disk while minimizing damage to the surface of the sample. In general we will refer to this stage as "dimpling" no matter how the thinning is achieved. Any damage you create at this stage will have to be removed during the final thinning process.

Most commercial mechanical dimplers use a small-radius tool to grind and polish the disk to a fixed radius of curvature in the center. Although the first instruments for dimpling were "home built" the commercial models (see Figure 10.4) are now well developed. You can control the load, precisely determine the thickness of removed material (the depth of the dimple), quickly change the polishing tool, and interrupt the process to remove the sample for closer examination before continuing. The investment is well justified for materials laboratories. One alternative that has been used successfully is a (recycled) dentist's drill and some imagination. Typically, dimpling can be carried out to produce regions ~10 μm thick, although in principle, precision dimpling with microprocessor control can sometimes produce electron transparent specimens which are <1 μm thick.

Figure 10.4. A dimpling apparatus showing grinding tool, and specimen support block.

As a general rule, the same guidelines apply as to all mechanical polishing; always gradually decrease the "grit" size and conclude with the finest available, again ensuring that the final specimen thinness is >2× the damage depth of the smallest grit dimension. The better the polished surface, the better the final specimen. If both sides of a disk are dimpled, the chances of final perforation occurring in the center are substantially increased, but in some cases you may wish to preserve one side of the specimen and thin from the other side only. One-sided dimpling is then essential prior to thinning to perforation.

Dimpling can also be performed chemically. Often in the case of Si this is achieved by allowing a jet of HF and HNO_3 to impinge (from below, as shown in Figure 10.5) on the Si disk which has the edges lacquered to produce a supporting rim. The HNO_3 oxidizes the Si and the HF removes the SiO_2. Similar approaches use Br and methanol for thinning GaAs. This dimpling method uses dangerous chemicals, but it is very efficient. It can even be carried to final perforation with care.

Anderson has revolutionized TEM specimen preparation through the development of the tripod polisher (Klepeis *et al.* 1988); this tool can help you thin your sample mechanically to less than 1 μm. You must consult the general references at the end of the chapter before using this tool. The tripod polisher, so called because it has three feet, is simply a device to hold your specimen while you mechanically thin it on a polishing wheel. A polisher can be purchased commercially or you can build your own.

> For some materials, such as Si, you can use this polisher to thin the specimen to electron transparency.

There are, however, several secrets in using the tripod polisher:

- You must use a very flat polishing wheel; the recommended approach is to use a glass platen. Take the greatest care in adjusting the micrometer to level the tripod.

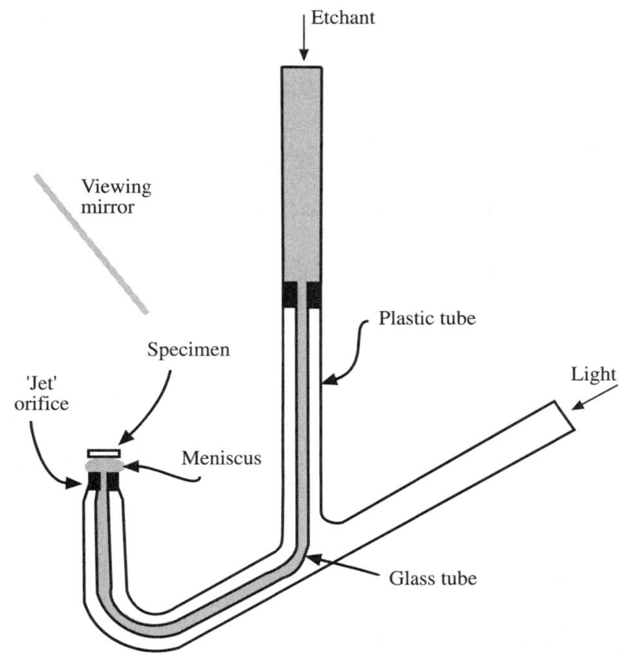

Figure 10.5. Surface dimpling using a chemical solution, e.g., to remove Si from one side of a disk. The light pipe permits visual detection of perforation using the mirror.

- You need a supply of fine diamond lapping films; these are not inexpensive but it is false economy to use them after they are worn. Always use a new sheet for polishing the second side of your sample since it is then particularly vulnerable.
- The diamond lapping films must not have an adhesive backing; you attach them to the glass platen by water tension and ensure that they are flat using a wiper blade. Bumps under the films will destroy your specimen.
- Any debris on the film will reduce its useful life; if the pad dries with polishing paste still present, you should discard it.
- Minimize the effect of debris, which you produce on the polishing film as you thin your sample, by paying careful attention to where you place the specimen on the polishing wheel; orient interfaces in cross-section samples normal to the radius and don't cross the debris trail.

With practice, you can dramatically reduce the time required for the final thinning step. This tool has had a major impact on making TEM a quality control instrument, particularly in the semiconductor industry.

10.4. FINAL THINNING OF THE DISK

10.4.A. Electropolishing

Electropolishing can only be used for electrically conducting samples such as metals and alloys. The method can be relatively quick (a few minutes to an hour or so) and it can produce foils with no mechanical damage. But it can change the surface chemistry of the specimen and it can be hazardous to your health, as you can see from the safety section at the start of the chapter.

The basic premise is that there is a certain applied voltage at which the current due to anodic dissolution of the specimen creates a polished surface rather than etching or pitting, as shown in Figure 10.6. The classical jet polish is shown in Figure 10.7A. By keeping the volume of the

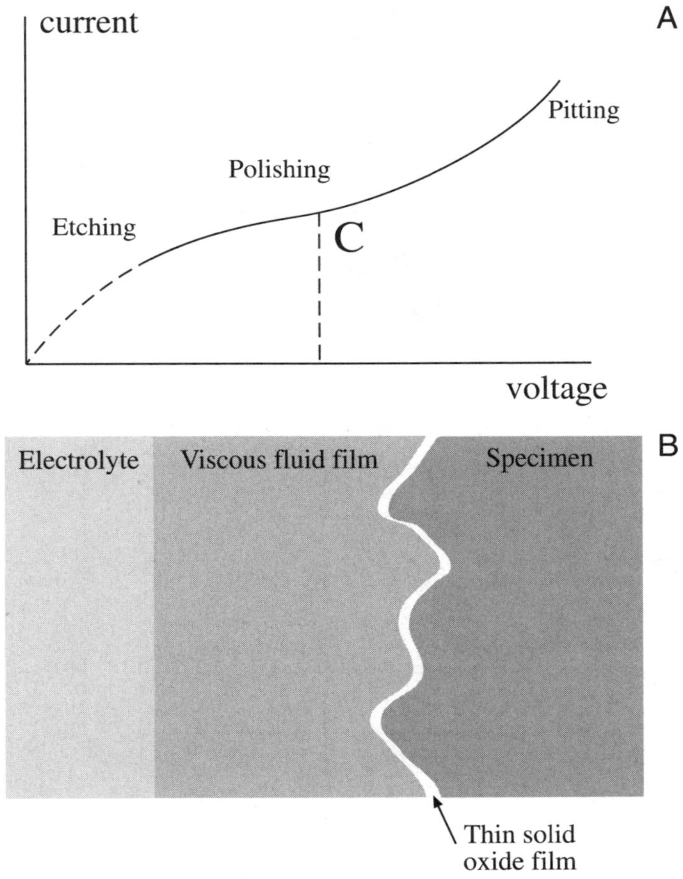

Figure 10.6. (A) Electropolishing curve showing the increase in current between the anode and the cathode as the applied voltage is increased. Polishing occurs on the plateau, etching at low voltages, and pitting at high voltages. (B) The ideal conditions for obtaining a polished surface require the formation of a viscous film between the electrolyte and the specimen surface.

reservoir constant, the jet falls under constant pressure. The voltage is applied between the tip of the pipette and the specimen. A twin-jet apparatus can be used to pump a jet of electrolyte onto both sides of the dimpled disk, as shown schematically in Figure 10.7B. A laser beam or light sensor detects transparency and a warning sound is given. At the warning, the electrolyte flow must be cut off immediately to prevent loss of thin area, and the disk must be rapidly extracted from the electrolyte and washed in solvent to remove any residual film of electrolyte which may etch the surface.

Electropolishing is a "black art." Undoubtedly you get better with practice, but reproducing the correct conditions of temperature, electrolyte solution chemistry, stirring rate, applied voltage, polishing current, etc., can only be achieved through trial and error.

10.4.B. Ion Milling

Ion milling involves bombarding your delicate thin TEM specimen with energetic ions or neutral atoms and sputtering material from your film until it is thin enough to be studied in the TEM. A schematic diagram is shown in Figure 10.8. The variables which you control include the voltage, temperature of the specimen [e.g., cold milling (liquid N_2)], the nature of the ion [Ar, He, or a reactive ion (iodine)], and the geometry (the angle of incidence).

An accelerating voltage of 4–6 keV is usually used. The ion beam will always penetrate the specimen to some extent, so we minimize this by inclining the incident ion

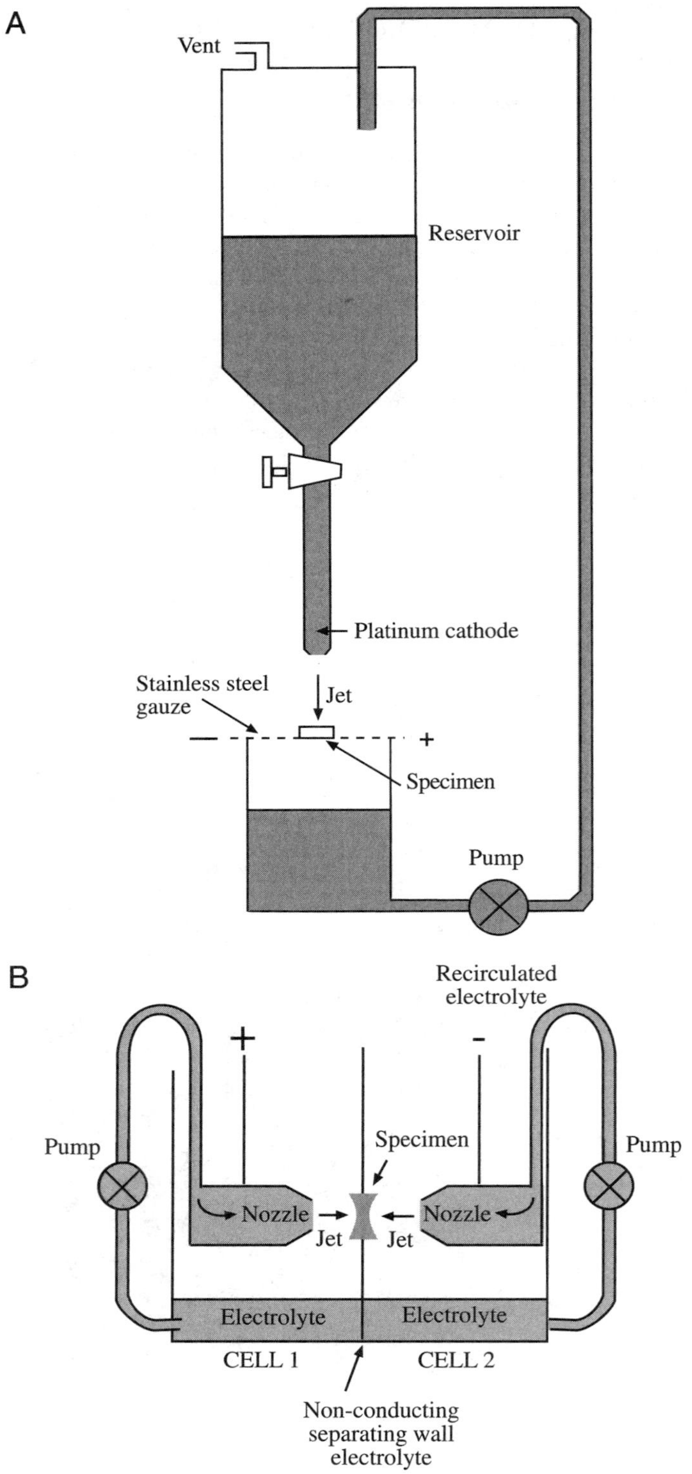

Figure 10.7. (A) Jet electropolishing by allowing a single jet of gravity-fed electrolyte to thin a disk supported on a positively charged gauze. The disk has to be rotated periodically. (B) Schematic of a twin-jet electropolishing apparatus. The positively charged specimen is held in a Teflon holder between the jets. A light pipe (not shown) detects perforation and terminates the polishing.

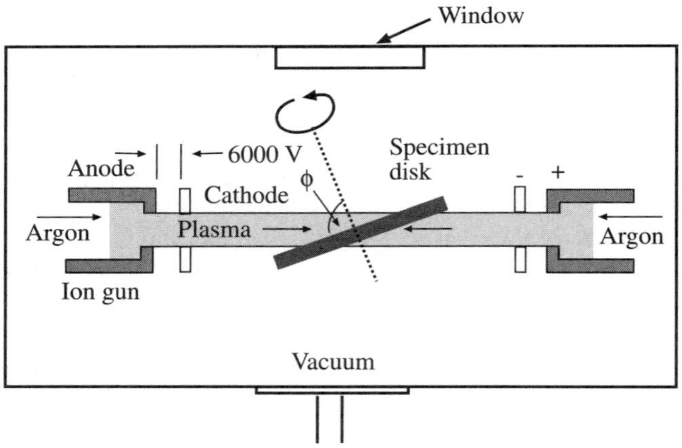

Figure 10.8. Schematic diagram of an ion-beam thinning device: Ar gas bleeds into an ionization chamber where a potential up to 6 keV creates a beam of Ar ions that impinge on a rotating specimen. Although not shown, the whole apparatus is under vacuum, the specimen may be cooled to liquid-N_2 temperatures, and perforation is detected by the penetration of ions through the specimen.

beam to the surface of the specimen. We usually align the ion beam at an angle of 15–25° to the surface. However, Barna (1992) has shown that this angle of incidence should be avoided in many cases since it leads to compositional thinning; use an inclination of ≤5° to avoid preferential thinning. Some implantation will occur so that the chemistry of the near-surface region is changed and the material is physically damaged (the top layer is often amorphized). If you use a low angle of incidence (<5°), you'll deposit the energy of the ion beam in a region close to the surface of the specimen. A lower beam energy or a lower *Z* ion will also do less damage, but in both cases milling time will increase. (Figure 10.9)

One thing you must remember is that ion thinning is closely related to ion-beam deposition. One manufacturer uses a similar arrangement to coat samples for SEM. The result is that material removed from one part of the sample can easily be redeposited elsewhere on the sample.

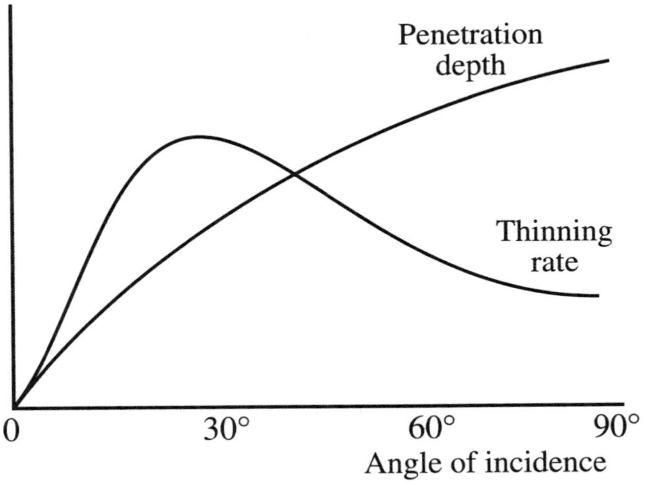

Figure 10.9. Variation in ion penetration depth and thinning rate with the angle of incidence. High incidence angles promote implantation, which is undesirable. The rate of thinning reaches a maximum at ~20° incidence, after which the beam penetrates rather than sputters the sample surface. Initial thinning should start at 20–30°, reducing to <5° as perforation approaches.

The theory of ion milling is complex. We can define the sputtering yield to be the number of atoms ejected per incident ion; the yield depends on the mass of the incoming ion. The yield also depends on the ion used and the sample being milled. The principal variables are:

- *Ion:* mass, energy, charge, and angle of incidence.
- *Target:* mass density, atomic mass, crystallinity, crystal structure, and orientation.

Ar is used because it is inert, heavy, and not naturally present in most samples. Special applications may use reactive iodine, or add oxygen, etc.; this idea of reactive-ion etching is commonly used in semiconductor processing. The problem is that the reactive ion may contaminate or corrode your thinning device, the diffusion pumps, etc. Heavy ions give less penetration, but create more damage.

Most of the thinning parameters are generally fixed except the ion energy, the angle of incidence, and any rotation of and the temperature of the specimen. A typical approach is to start with rapid thinning conditions (heavy ions, high incidence angle) and slow the thinning rate as perforation approaches. The effect of incidence angle on the thinning process is shown in Figure 10.9. Cooling the specimen is recommended for almost all materials; otherwise, it is possible that the ion beam might heat it to 200°C or higher. Even in metals which have good thermal conductivity, the creation of vacancies through ion damage can cause diffusional changes equivalent to heat treatment at such temperatures.

You may encounter discussions of whether to use ions or neutral atoms; one idea is that neutralized ions should not be affected by charging of a ceramic specimen. It is not clear that neutral atoms remain neutral throughout the thinning process, so this may be a moot point.

Ion milling is the most versatile thinning process, being used for ceramics, composites, polyphase semiconductors and alloys, and many cross-section specimens. In addition, fibers and powders, which constitute a wide range of important materials, can also be thinned by ion milling. To do this, you have first to embed the particles or fibers in epoxy, and transfer the mixture into a 3-mm brass tube for strength. The next step is to saw the tube/epoxy mixture into 3-mm disks and finally dimple and ion mill to electron transparency, as shown in Figure 10.10. A similar method (but without the brass tube) can be used prior to ultramicrotomy of powders and fibers (see Section 10.6.B).

Remember: Always beware of artifacts; some stories best illustrate this. Goodhew (1985) reports that Ar bubbles form in silicon at a depth of ~10 nm after 5-keV thinning. Chemical analysis (EDS) of some β-aluminas which had the correct structure by HRTEM (composition $K_2O{\cdot}11Al_2O_3$) gave a composition with the K completely replaced by Ar (glasses and zeolites can accommodate large amounts of Ar). Cooling the specimen can often reduce contamination and surface damage. It is best to use two ion guns. If this is not acceptable, because you want to study the surface region, then you may want to coat one side with a polymer protective lacquer, and then dissolve this coating after thinning to remove sputtered material.

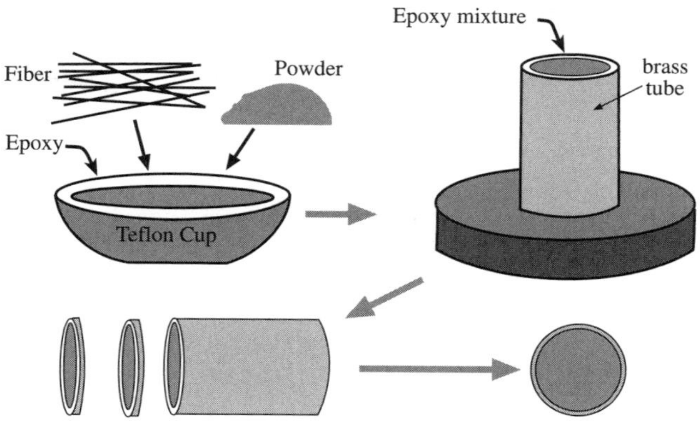

Figure 10.10. Sequence of steps for thinning particles and fibers by first embedding them in epoxy and forcing the epoxy into a 3 mm (outside) diameter brass tube prior to curing the epoxy. The tube and epoxy are then sectioned into disks with a diamond saw, dimpled, and ion milled to transparency.

Why Rotate and Cool the Specimen?

The specimen is usually rotated (at a few rpm) during thinning, otherwise you tend to get surface structure—grooves which run in certain directions; if you see these, check to see that the rotation has not stopped. In the preparation of cross-section specimens, you may use beam blockers and rotation control; in the first you physically block the sample so that it cannot thin, say at an interface, preferentially. In the second, you vary the rate at which you rotate the sample to achieve the same effect. The latter is preferred if it is available, since the time spent thinning the specimen is maximized.

Why cool the specimen? You can minimize atom migration in or on the specimen. We noted above that the specimen might be heated to >200°C otherwise. An additional advantage is that the cooling system also cools the surroundings to give a contribution of cryopumping and simple cryotrapping. However, you have to give the specimen time to warm up after milling, which can increase preparation times.

Tilting the Specimen

This depends on your ion miller, but if you're choosing a new machine there may be an advantage in tilting the gun rather than the specimen. If the specimen is inclined, then you need a clamping ring and you may sputter this when you thin the specimen. This has led to the development of ion polishing instruments (see later) where the ion thinner has been optimized to provide a low angle without a retaining clamp. The specimen rests on a support and can be thinned at an angle of 4–5°.

Practical Design of the Ion Miller

The schematic diagram in Figure 10.8 doesn't do justice to a modern ion miller, which is a highly sophisticated piece of equipment. Two ion guns are available to thin from each side. The operating vacuum is $<10^{-5}$ Torr without Ar and 10^{-2}–10^{-3} Torr when Ar is bled into the gun. The ion guns are basically hollow chambers into which the Ar is introduced; then it is ionized and accelerated through a hole in the cathode. The hole gradually enlarges due to ion sputtering and cathodes need replacing after some time to maintain a high-intensity ion beam. More advanced gun designs incorporate saddle fields to focus the ion beam at the specimen and increase the thinning rate. The beam can be neutralized in some systems if the charged ions cause too much damage.

Some special phrases you'll encounter:

- Reactive ion milling. The classic example is the use of iodine in the work described by Cullis and Chew (1988). Iodine has a clear advantage for InP, where In island formation under Ar thinning is suppressed. In CdTe only growth defects were observed in iodine-thinned specimens, but many other defects were found in the same material thinned using argon ions (Figure 10.11).
- The PIMS. Gatan's precision ion milling system provides a built-in ion microscope to view the specimen (through SE emission) so that you can then choose a particular area to thin. This is only useful for prethinned specimens where you can locate a very small area (~1 μm × 1 μm) that needs further thinning.
- Beam blockers and variable rotation speeds. Often the epoxy in a cross-section specimen thins faster than the specimen. Therefore, we want to direct the ion beam at the different materials for different amounts of time. The two approaches used are blocking the beam geometrically using "beam blockers" or varying the rotation velocity; e.g., you don't want the beam to thin along the interface. The latter approach can be extended further to oscillate the specimen, always keeping the ion beam at the same angle of incidence, so that it is never parallel to the interface.
- The PIPS. Gatan's precision ion polishing system combines high-powered ion guns and a low angle of incidence (4°) to thin one side of a specimen with minimum surface damage and heating. The low incidence angle removes any surface roughness and differential thinning

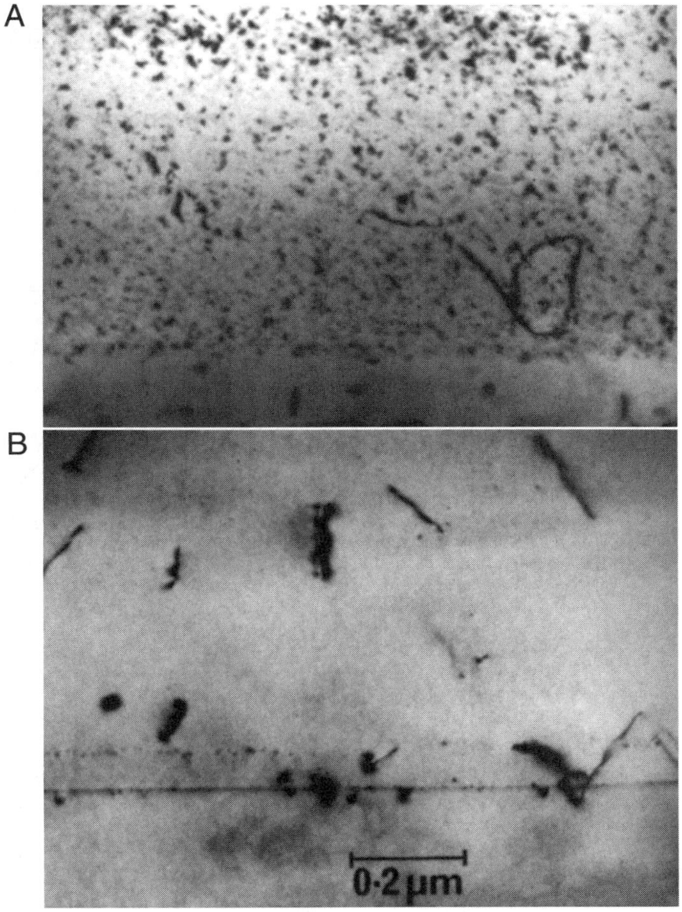

Figure 10.11. BF images of CdTe showing (A) defects (dark spots) in Ar-thinned specimen and (B) undamaged crystal thinned by reactive-iodine ion milling. The residual defects in (B) were formed during CdTe crystal growth.

problems, while the high-power guns ensure reasonable thinning rates (Alani and Swann 1992).

Some final points to remember:

- Materials thin at different rates. It's a good idea for the person responsible for the ion millers to run a test specimen periodically with nominally the same conditions, to be sure that the machine is still working optimally.
- Don't start with a thick sample. Always make the surface as smooth as possible before beginning to ion thin.
- Keep a record of what conditions you use: record the beam current, angle of incidence, rotation rate, and kV.
- Ion milling will form a layer on the surface which will probably be a combination of amorphous, highly damaged, and implanted material! The chemistry of the layer will be different than the rest of the specimen. The thickness of crystalline material will thus be less than the total thickness.

10.5. CROSS-SECTION SPECIMENS

The cross-section specimen is a special type of self-supporting disk. You must master this preparation technique if you are studying interfaces. We have often stressed that one of the principal limitations of the TEM is its insensitivity to variations in the structure and chemistry of the specimen in the direction of the electron beam. Therefore, if we are to look at structural and chemical variations close to an interface we have to prepare specimens in which the interface is parallel to the electron beam and this involves cross-sectioning the sample. The most widely used cross-section samples are semiconductor devices, which often have multiple layers and therefore have multiple interfaces. But any composite materials, samples with surface layers (e.g., oxide–metal interfaces), MBE specimens, quantum-well heterostructures, etc., are candidates for this type of preparation.

There are numerous techniques for preparing cross-section specimens and many details are reported in four Materials Research Society proceedings (Bravman *et al.* 1988; Anderson 1989, 1990; Anderson *et al.* 1992), so we'll only describe a few basic principles. First, rather than trying to thin one interface only, the sample can be cut and glued together to produce several layers, rather like a club sandwich. Then the sandwich is sectioned such that we can see the layers, as shown schematically in Figure 10.12. In this process, a critical step is the gluing of the sections to form the sandwich. Several epoxies are available that cure at low temperatures, so that you won't heat treat the specimen inadvertently. The thickness of the epoxy layer must be such that it is thick enough for good adhesion, but not so thick that it is completely thinned away during final ion milling.

You can then cut the glued sections into 3-mm rods using an ultrasonic drill. Alternatively, you can cut the samples smaller and encase them in a 3-mm thin-walled tube. Section the tube into disks, which you can then ion thin. The advantage of this method is that the final specimen has a thick ring of the metal tube around it, which gives it mechanical stability. With multiple interfaces the final thinning is almost always guaranteed to produce electron transparency at a useful region.

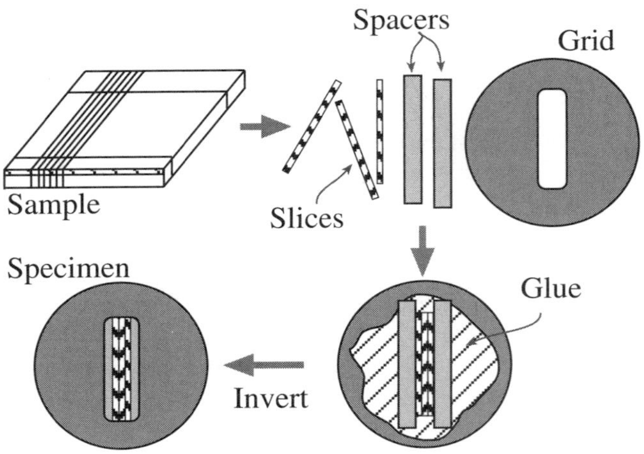

Figure 10.12. Schematic sequence for cross-section specimen preparation; the sample is cut into thin slices normal to the interfaces which are glued together between spacers which could be Si, glass, or some other inexpensive material so that they are wider than the slot in the grid. The "club" sandwich is then itself glued to the grid (over the slot) and ion milled to perforation.

10.6. SPECIMENS ON GRIDS/WASHERS

The alternative to self-supporting disks is to make small electron transparent portions of the specimen, or create particles and support them on a thin film on a grid or washer. We can deposit these small particles on amorphous or crystalline films. The classic example is the amorphous carbon film: the holey carbon film. Some of the particles of the material of interest will be located partially over a hole so that they do not overlap anything else.

The thin supporting film should have a uniform thickness; the idea is that you are not actually interested in this material and therefore want to minimize its effect on the image of the material you are interested in.

The particles may stick to the film or may have to be clamped between two grids. Special hinged "oyster" grids (see Figure 10.2) are available which make this very easy. Some of the processes we've already discussed can be used to make these specimens.

10.6.A. Electropolishing—The Window Method for Metals and Alloys

Electropolishing is an application of electrochemistry and is regarded by many as a "black art": a recipe which works one day but might not work the next. We can electropolish a thin sheet of metal. First, cut the sheet into a ~1 cm × 1 cm square then seal the edges with a polymer lacquer to prevent preferential attack. The "window" of exposed metal is immersed in electrolyte (usually cooled to slow the rate of dissolution), surrounded by a cathode and a voltage is applied, as in Figure 10.13A. The solution may or may not be stirred. The correct voltage will ensure that a viscous layer of electrolyte builds up at the surface of the specimen which results in uniform controlled thinning without pitting or corrosion. After some time, which you have to determine experimentally, the sheet is removed, cleaned, and turned through 180° and replaced in the bath, as shown schematically in Figure 10.13B. If this procedure is done correctly (and this might require several rotations)

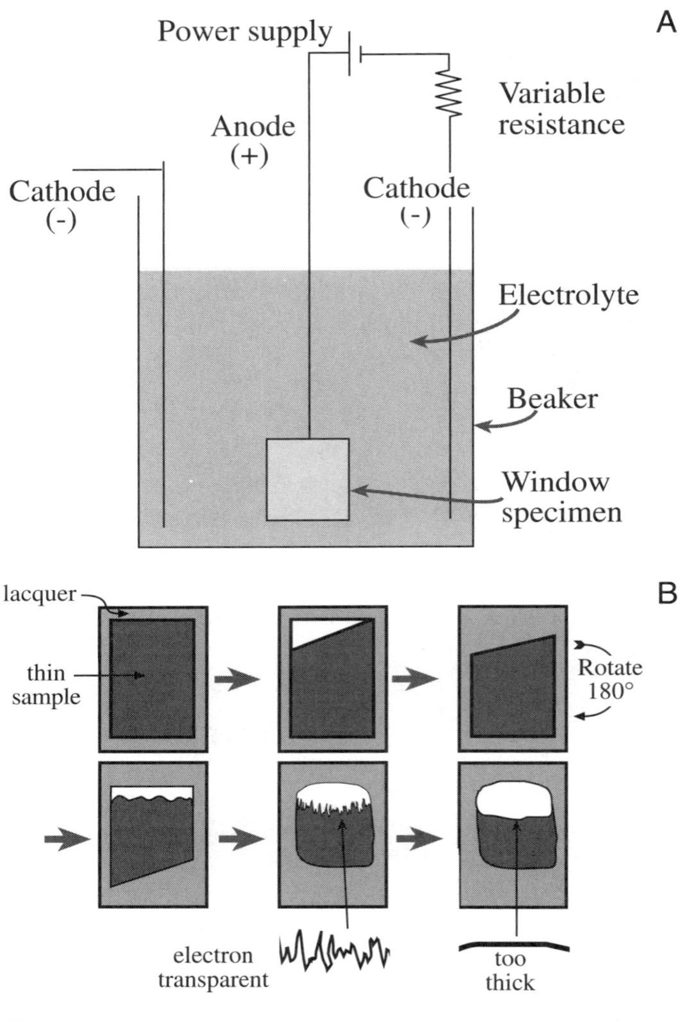

Figure 10.13. Window polishing: (A) A sheet of the metal ~1 cm^2 is lacquered around the edges and made the anode of an electrolytic cell. (B) Progress during thinning: the initial perforation usually occurs at the top of the sheet; lacquer is used to cover the initial perforation and the sheet is rotated 180° and thinning continues to ensure that final thinning occurs near the center of the sheet; if the final edge is smooth rather than jagged it is probably too thick.

the sheet will finally thin in the center. If final thinning occurs too near the top of the sheet, the edge of the perforation is smooth and relatively thick. After perforation, remove the sheet and cut off slivers of material from around the perforation using a scalpel under an inert solvent such as ethanol. Catch the floating slivers on oyster grids, dry them, and they are ready for viewing.

10.6.B. Ultramicrotomy

The microtome has long been used for sectioning biological materials. (A tome is a "piece cut off" but microtome refers to the instrument used to cut a very thin tome, unlike the one you're reading.) With care and much practice the biologist can reconstruct a 3D picture of the specimen. For visible-light microscopy the specimens are usually <0.1 mm thick; for the TEM the slices may be <100 nm thick and the instrument is known as an ultramicrotome. These instruments are routinely used for biology or for polymers where the samples tend to be quite soft (Sawyer and Grubb 1987). More recently they have been used for many studies of crystalline materials (Malis 1989). The principal advantages of the technique are that it leaves the chemistry unchanged and is thus ideal for AEM specimens, and you can use it to create uniform thin films of multiphase material. The main disadvantage, of course, is that it introduces a deformation structure to the materials and therefore is most useful in cases where the defect structure is of secondary importance.

The ultramicrotome operates by moving the specimen past a knife blade. The blade can be glass (cheap) for soft materials but will be diamond for harder ones. Since there are so many possible applications, we will describe a few and refer to the references at the end of the chapter for more details. Two processes can occur in principle: the knife can cut the sample if it is soft, or the knife can cause a partly controlled fracture if it is hard/brittle. In either case the limiting process is usually plastic deformation of the specimen. The principles of this technique are shown in Figure 10.14.

You will also find ultramicrotomy useful if you want to study particles or fibers which are too small to thin individually but are too large to be electron transparent. You can embed the sample as we saw for the ion-thinned particles, but without using the metal sheath (see Figure 10.10). We also use epoxy if the sample contains so many interconnected pores that it cannot be thinned mechanically. In this case, place the sample in a vacuum chamber, pump out the chamber, and coat the sample with epoxy using a dropper in the chamber. When the sample is fully encapsulated, admit air to the chamber so as to push the epoxy into the pores. After curing, you can ultramicrotome the sample in the usual way.

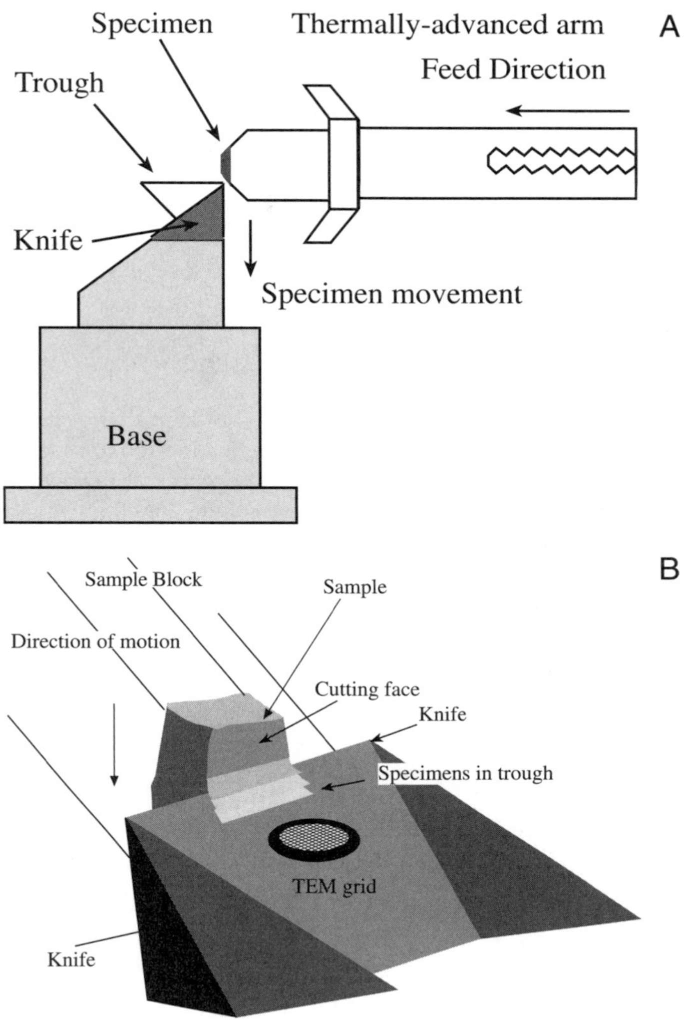

Figure 10.14. Ultramicrotomy: (A) The sample is first embedded in epoxy or some other medium, or the whole sample is clamped and moved across a knife edge. (B) The thin flakes float off onto water or an appropriate inert medium, from where they are collected on grids.

10.6.C. Grinding and Crushing

Many brittle materials such as ceramics and minerals are most easily prepared by crushing in a clean pestle and mortar (preferably in an inert liquid). The liquid containing the particles can then be ultrasonically stirred and allowed to settle. The thin particles are too small to be seen and the supernatant liquid in which they remain should appear clear. A drop of this liquid, if placed on a holey carbon film on a grid, will evaporate in a dry environment, leaving a distribution of the particles on the support film. If the particles have to be crushed dry, then agglomeration can be a problem. Electrostatic forces sometimes cause small particles to clump together and distributing them on a grid can be very difficult. In these cases, it sometimes pays to mix up the crushed material in an epoxy, then ultramicrotome the epoxy, as we just described in the previous section.

Dust particles found in airborne pollution can be collected simply by leaving the grid and a support film out for a period of time.

10.6.D. Replication and Extraction

These methods are among the oldest TEM specimen preparation techniques. We use direct replication to study fracture surfaces or surface topography in general. Evaporate a carbon film on the surface of interest, then etch away the underlying surface with an acid so that the carbon film floats off. If you coat this film with a heavy metal at an oblique angle, you will enhance mass-thickness contrast (see Chapter 22); support the film on a grid for observation. As an alternative (Figure 10.15A) you can first replicate the surface by softening a plastic, pressing it on the surface, and allowing it to harden. Pull off the plastic replica, coat it with carbon, then dissolve the plastic with a suitable solvent and pick up the carbon replica on a support grid. If the carbon replica is directly from a metal surface, it may be necessary to dissolve some of the metal with acid then float off the carbon onto distilled water before picking up on a grid, as shown in Figure 10.15B. After picking up on a grid it may be useful to coat the replica obliquely with a heavy metal to enhance any topographic (thickness) contrast.

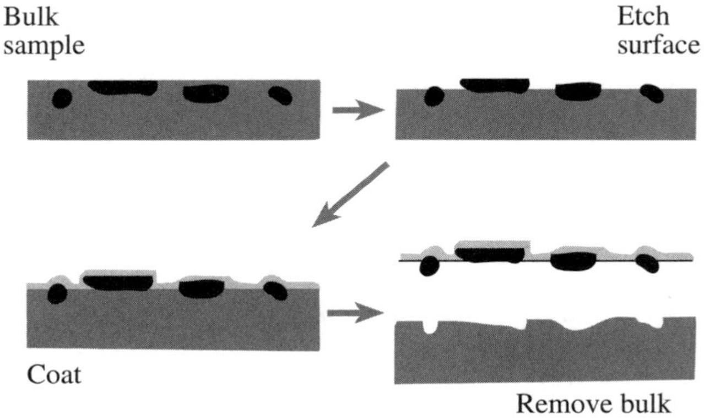

Figure 10.16. Extraction replication: particles embedded in a matrix are revealed by etching the matrix, which leaves the particles standing proud of the surface; a thin amorphous carbon film is evaporated over the particles, then the rest of the matrix is etched away leaving the particles adhering to the carbon film.

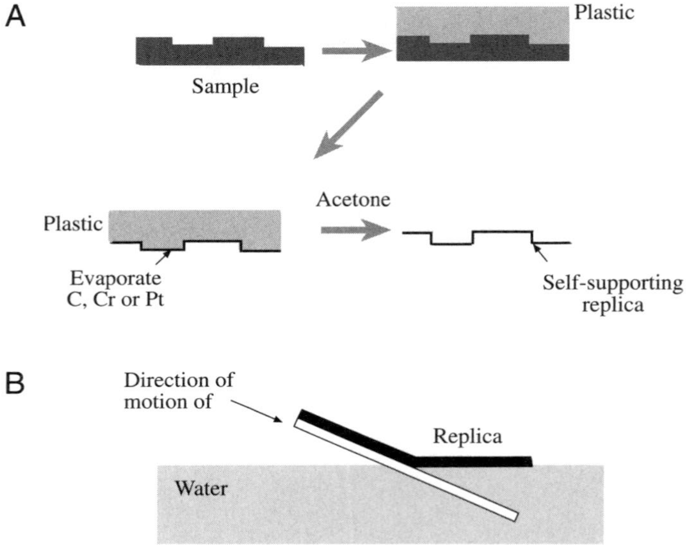

Figure 10.15. (A) Replication of a surface by the two-step method: spray acetone on the surface to be replicated before pressing a plastic (usually cellulose acetate) onto the surface which softens in contact with the acetone; the plastic is removed from the surface when it has hardened and a C, Cr, or Pt film is evaporated onto the replicated plastic surface; the plastic is then dissolved with acetone and the evaporated film retains the original topography. (B) Alternatively, the direct carbon replica of a metal surface may be floated off on distilled water after scratching the carbon and etching to free the film, which may subsequently be shadowed obliquely to enhance the topography.

Extraction replication has seen a resurgence of interest since AEM techniques appeared, because we can extract a particle from its surrounding matrix, thus allowing us to analyze that phase alone without interference from electron scattering into the matrix.

The various steps for extraction are shown in Figure 10.16. The sample is polished metallographically to expose the particles on the surface. An appropriate etching process is used to remove the matrix such that the particles stand proud of the surface. A carbon film is evaporated onto the surface and scored into ~2-mm squares. Then the etching is continued. As the matrix is dissolved, the squares of carbon film float to the surface carrying the particles with them. Catch one of these squares on a grid and you have your specimen. Again, oblique shadowing may be useful to enhance image contrast, but not if AEM is to be used.

10.6.E. Cleaving

This is one of the oldest techniques and has been used to make thin specimens of graphite, mica, and other layer materials which are weakly bonded along one plane. The idea is to attach adhesive tape to both sides of the sample and then pull the two pieces of tape apart. This process is repeated until the specimen is thin enough for TEM. You can really only tell this by experience: as it becomes thinner, graphite becomes a lighter shade of gray in transmitted visible light. Molybdenite (MoS_2) becomes a lighter shade of green. Place the tape with the thin flake of material in a solvent to dissolve the glue (all traces of glue *must* be removed). This technique is not as easy as it once was. The glues used to be readily soluble in trichlorethylene, which is now a known carcinogen.

10.6.F. The 90° Wedge

The 90°-wedge specimen was developed because many compound semiconductors such as GaAs are grown with a (001) surface and can be easily cleaved on the (110) and ($1\bar{1}0$) planes, which are perpendicular to this growth surface. When you are practiced at cleaving the sample as shown in Figure 10.17, you can examine a specimen in the TEM within 30 minutes of completing the growth.

Mount the specimen as shown in the figure, preferably so that you won't need to tilt it in the microscope. Although the specimen is only transparent close to the edge of the "hole," you will have a long strip of material suitable for viewing. As always, beware of artifacts. If your specimen is perfect, you will know exactly how thick it is at the position you choose for study. We will find this wedge useful when we discuss image contrast in Part III.

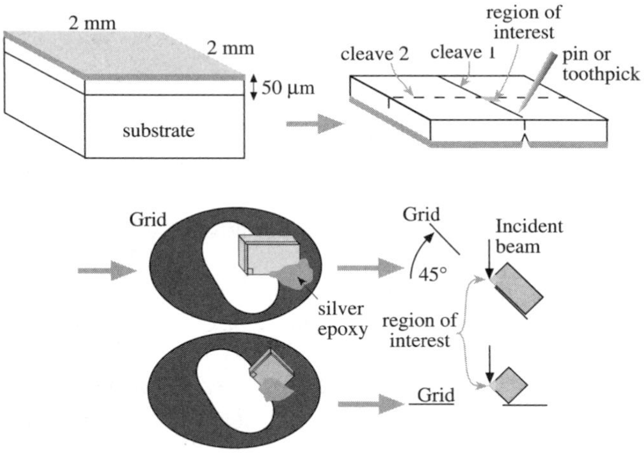

Figure 10.17. The 90°-wedge specimen: prethin to create a 2-mm square of the multilayers on a Si substrate; scribe the Si through the surface layers, turn over, and cleave; inspect to make sure the cleavage is clean, giving a sharp 90° edge; reject if not; mount the 90° corner over the edge of a hole in a Cu grid, then insert in the TEM; note that two different orientations are available from a single cleavage operation.

10.6.G. Lithography

Using a technique developed for advanced engineering applications, lithography is used in the microelectronics industry to define fine lines of width down to 100 nm. An illustration of how lithography can be used specifically to prepare TEM specimens (as opposed to generating a structure which might best be characterized by TEM) is shown in Figure 10.18 (Brown and Sheng 1988). We can draw lines on the layered material using standard lithographic techniques. Material on either side of the lines is then removed by etching (chemical or ion) to give a plateau which is thin in one direction. We then remove most of the remaining substrate and attach the specimen to a support washer. We can then observe the specimen directly in the TEM. Although the width (formerly height) of the electron transparent region is narrow, it can extend across the entire hole in the 3-mm disk. The major disadvantages or limitations of the technique are: (i) the dimension in the direction of the electron beam is fixed by the lithographic capabilities and (ii) tilting the specimen may quickly cause the thicker region to block the electron beam.

10.6.H. Preferential Chemical Etching

The principle behind this technique is the same as for lithography: we remove part of the sample to leave an area which is electron transparent. The trick is to keep part of the final specimen thick enough for handling, or ideally for supporting, the specimen. Naturally, this approach only works with certain materials although the principle might be extended to other thin films. The technique has been used for III-V compounds where $Al_{1-x}Ga_xAs$ acts as an etch stop for GaAs, and for Si where an etch stop can be pro-

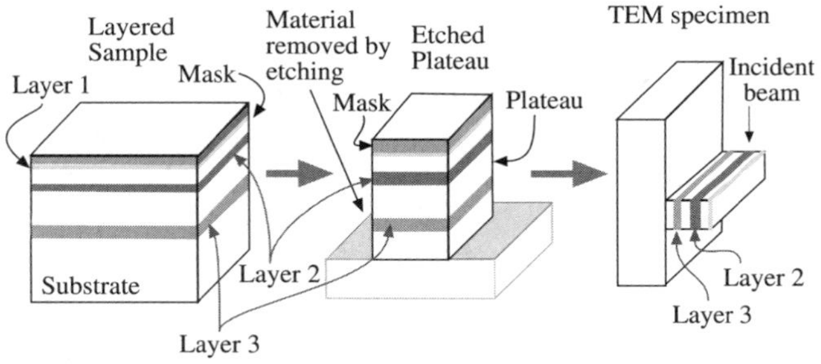

Figure 10.18. Etching of a multilayer sample. Etch away most of the sample, leaving a small etched plateau; mask a region < 50 nm across and etch away the majority of the surrounding plateau. If this thin region is turned 90° and mounted in a specimen holder, the interfaces are now parallel to the electron beam.

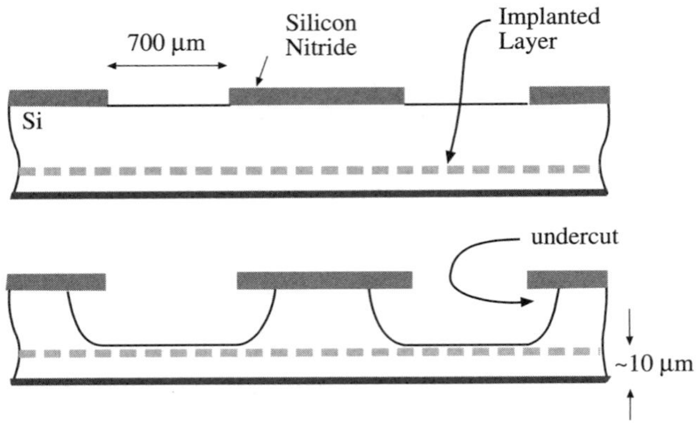

Figure 10.19. Lithographic techniques applied to thinning a multilayer specimen: in the top diagram, the unthinned sample is shown with a grid of Si_3N_4 barrier layers evident. Etching between the barrier layers, shown in the lower figure, produces an undercutting down to the implanted layer (e.g., B) which acts as an etch stop, producing a uniform layer ~10 μm thick. Further thinning with a different solution produces large areas of uniformly thin material (not shown) supported by the Si_3N_4 grid and the remaining unthinned regions.

duced by implanting with boron (Figure 10.19). In both cases, the resulting thin layers may find use as substrate materials for thin-film studies rather than as the subject of study in their own right.

10.7. STORING SPECIMENS

The best advice is to look at your specimen as soon as possible after preparation. If that is not possible, then keep the specimens under optimum conditions. Usually this means keeping them dry (water affects most materials), perhaps in an inert atmosphere (dry nitrogen works well, or a dry-pumped vacuum desiccator) and in an inert container (a petri dish with filter paper).

The next problem is long-term storage; for periods up to 1 month, you can use the above procedure. If you want to keep the specimen longer your choices can be more difficult. Don't use gelatin capsules for anything resembling "delicate" material. Don't use slotted grid-holders for anything which might deform (break or bend) during handling; that rules out self-supporting ceramics, metals, and semiconductors. Always use vacuum tweezers to manipulate delicate specimens. Remember, your most important specimen is the one most likely to break, bend, interact with sharp tweezers, or jump onto the floor.

Lastly, old specimens can be cleaned by ion polishing. This process does thin the specimen further, so you may lose the area you originally studied. Ion polishing can also be useful for "sectioning" specimens.

CHAPTER SUMMARY

Specimen preparation is a craft and there is no substitute for hard work and careful, detailed experimentation as you seek to master it. This is the most tedious aspect of all TEM work but, if you invest the time, your reward will be the best of times on the TEM itself. The quality of your data is at least directly proportional to the quality of your specimen (and this relationship is often far stronger than the linear nature just implied). You simply have to find the method that works best for your particular material. While there are many cookbooks available, the recipes are often too individualized and not to your specific taste.

There are few rules for specimen preparation except that thinner is usually better, although such specimens are more prone to artifacts. Think about each step and what it might do to change the microstructure or microchemistry of your material. Take care to avoid the physical dangers that are present whenever you use dangerous chemicals, ionizing radiation, or sharp knives. Be clean, use fresh materials, tidy up after yourself, and apply all the other lessons that you learned in kindergarten!

Although all the equipment mentioned here is available commercially, most was originally developed on a shoestring budget in someone's lab so you can always build your own electropolisher or even an ion mill. If you are working with brittle materials, buy or build a tripod polisher and learn how to use it.

We stress once again that you must know what you want to study in your specimen before you begin specimen preparation. Figure 10.20 is a useful flow chart (Goodhew 1988) which summarizes the various possible options. Be aware of the limitations of the method you choose, particularly the artifacts introduced. In this respect, Table 10.1 (Malis 1989) is a nice summary of the artifacts introduced by various methods.

A last reminder: The recipe books listed below are a great source of ideas. New recipes are appearing all the time. As is often the case in cooking, it helps to see an expert chef in action to realize what is possible. In other words, when you have seen a really good TEM specimen, you'll know what yours should look like.